RETROFITTING DESIGN OF BUILDING STRUCTURES

RETROFITTING DESIGN OF BUILDING STRUCTURES

EDITED BY
XILIN LU

Science Press

CRC Press
Taylor & Francis Group
Boca Raton   London   New York

CRC Press is an imprint of the
Taylor & Francis Group, an **informa** business

CRC Press
Taylor & Francis Group
6000 Broken Sound Parkway NW, Suite 300
Boca Raton, FL 33487-2742

First issued in paperback 2017

ISBN-13: 978-1-4200-9178-6 (hbk)
ISBN-13: 978-1-138-11322-0 (pbk)

Library of Congress Cataloging-in-Publication Data

Retrofitting design of building structures / editor, Xilin Lu.
    p. cm.
    "A CRC title."
    Includes bibliographical references and index.
    ISBN 978-1-4200-9178-6 (hard back : alk. paper)
    1. Buildings--Remodeling for other use. 2. Structural analysis (Engineering) I. Lu, Xilin.

TH3411.R485 2010
690'.24--dc22

    2009049458

Visit the Taylor & Francis Web site at
http://www.taylorandfrancis.com

and the CRC Press Web site at
http://www.crcpress.com

Contents

Preface

Retrofitting of building structures, including maintenance, rehabilitation and strengthening, is not only an important issue in urban construction and management, but also a frequent problem to structural engineers in property management disciplines. This book is based on the work of the authors who have carried out various structural retrofitting practices, followed the basic principles of structural analysis and design, and innovatively used various structural codes for design, assessment and retrofitting of building structures using newly developed technologies worldwide.

Beginning with the procedure of structural retrofitting, this book gradually introduces the significance of structural retrofitting; the inspection methods for structural materials, structural deformation and damages; retrofitting design methods and construction requirements of various structural systems; and some practical examples of structural retrofitting design and construction. In the introduction of various practical examples, emphasis is put not only upon conceptual design, but also on constructional procedure design, so that a structural retrofitting design work should be completed by both structural analysis and detailed constructional measures.

This book may be used by the professionals who understand the structural design concept and have basic knowledge of structural materials, structural mechanics and construction technology. This book may also be used as reference material for the teachers and students in civil engineering institutes.

The first chapter and some of the second chapter were written by Xilin Lu, and the rest of chapters of the Chinese version were originally written by Guofang Jin, Siming Li, and Deyuan Zhou. The final manuscript was summarized and edited by Xilin Lu. All the figures in the text were plotted by graduate students: Zhen Yang, Yonghui Ruan, Liming Xiang, and Cunzhong Liang.

The English version of this book was translated by a group of graduate students: Linzhi Chen, Zhenghua Cai, Chunjie Gan, Kai Hu, Zheng Lu, Jing Kang, Chongen Xu, Fuwen Zhang, Jie Zhang, and Xudong Zhu, at the Research Institute of Structural Engineering and Disaster Reduction at Tongji University under the guidance of Ying Zhou, and the figures were drawn by Yi Bo, Juhua Yang, and Xueping Yang. The final manuscript was edited by Xilin Lu with the assistance of Ying Zhou.

The authors of this book sincerely appreciate the work done by the above-mentioned students who spent their summer holidays completing the translation and editorial work of this book.

The author is grateful for the financial support in part from the National Basic Research of China (Grant No. 2007CB714202) and National Key Technology R&D Program (Grant No. 2009BAJ28B02).

CHAPTER 1

Introduction

1.1 The Significance of Structural Retrofitting

Before the 1980s, the key issue for retrofitting of building structures was the assessment and strengthening of old houses. After the middle of the 1990s, however, structural assessment and strengthening for both old and new buildings is becoming more intensive due to the rapid development of the construction industry in China. Especially in the real estate market, privately owned houses have become a majority, and many disputes between house developer and house buyer have resulted. To solve these problems technically it is necessary to perform structural assessment and strengthening of building structures. In some cases the result of inspection and assessment of building structures is becoming an increasingly important basis for government officials to mediate the disputes in the real estate market.

Generally speaking, the following building structures need to be evaluated and retrofitted.

(1) The buildings whose serviceability or strength cannot meet the requirements of structural codes or regulations, due to misuse, irregular maintenance, aging of materials and structures.

(2) The buildings that have quality or safety problems due to design flaws or deficiency in construction quality. These problems are often met in new construction and existing buildings.

(3) The buildings in which structural damages are caused by disasters such as earthquakes, strong winds, fires, etc.

(4) Those historic buildings and memorial buildings that need to be rehabilitated and protected.

(5) The buildings that will be reconstructed, or have additional stories built.

(6) The buildings whose structural members may be changed during renovation, which may influence the performance of whole structural system.

(7) When the buildings are located close to the site of a deep pit foundation of a new construction, this deep excavation may cause unequal settlement of the surrounding soil and the surrounding buildings may consequently face potential damages or risks. The assessment and retrofitting of this kind of building is also an important safety measure for the construction of the deep foundation as well as the new structure.

1.2 The Retrofitting Procedure of Building Structures

The retrofitting procedure of a building structure is as follows.

(1) Inspection of mechanical properties of structural materials

Generally the material properties may be obtained from design documents and construction daily records, especially the checking and accepting record upon completion of the project. If there is doubt concerning material strength, testing of materials is necessary.

(2) Assessment of structural vulnerability and safety

Vulnerability assessment is to evaluate serviceability of a building structure by inspecting the appearance of members and structural system to provide a basis for maintenance and rehabilitation of the building. The safety assessment evaluates the strength of members and structural system by structural analysis and section checking according to related design code to provide a basis for structural retrofitting of the building.

(3) Working out the retrofitting scheme

After completing the assessment of structural vulnerability and safety, a detailed retrofitting plan can be laid out by comprehensively considering the service requirements of the building, future life-cycle of the building, construction condition, cost issue, etc.

(4) Design of retrofitting construction drawings

Usually the construction drawings can be done according to the retrofitting scheme together with detailed attention paid to connections between new structure and existing structure, as well as the safety issue of existing structure during the retrofitting construction phase.

(5) Inspection during construction and acceptance checking after retrofitting

Designers and inspectors are required to go to the construction site to solve various problems that have occurred. Especially when the existing structure does not coincide with the construction drawings, the designers must go to site and modify their design drawings accordingly. The acceptance checking after retrofitting is as important as it is for new construction. In some cases measurement after retrofitting is also necessary, especially for very important and large-scale construction.

1.3   The Principles of Retrofitting Design for Building Structures

The principles of retrofitting design for building structures are as follows.

(1) Strengthening of members versus strengthening of structural system

There is no doubt that the members that do not meet safety requirements must be strengthened. However, there is often an underlying mistake that the strengthening of whole structural system is neglected. For example, strengthening of individual members may result in redistribution of strength and stiffness in the structural system, therefore strengthening of the whole structural system must be considered. Another important issue is that strengthening of connection between members is quite influential to structural integrity.

(2) Local strengthening versus global strengthening

Local strengthening of an individual member can be carried out only if the strengthening does not affect the structural performance of the whole system. For example, if there is a local damage to beam and slab due to an equipment explosion within a small area of a building, then strengthening of the damaged beam and slab is enough. When lateral strength of a structural system is too weak to meet lateral deformation requirements under earthquake action, strengthening of the whole structural system is necessary.

(3) Temporary strengthening versus permanent strengthening

The standards and requirements for temporary strengthening may be lower than those for permanent strengthening.

(4) Special considerations for earthquake resistant strengthening.

a. The distribution of strength and stiffness along structural height should be uniform.

b. Vertical structural member should be continuous so as to transfer loading smoothly.

c. Earthquake action may be increased due to the change of natural characteristics of structure by strengthening the structural system.

d. The torsional response of whole structure should be reduced whether adding new structural members or strengthening existing members.

e. Detailed construction measures should be taken to any weak parts of a structure.

f. To ensure the structural system to be more resilient, so as to prevent brittle failure of members, and to eliminate the poor earthquake resistant mechanism such as "strong beam weak column" and "strong member weak joint".

g. To consider site responses of structures according to the specific condition of the construction site. Response of the strengthened structure must be controlled to be smaller than that before strengthening. According to earthquake damage and basic theory of seismic analysis, the seismic response of stiff structures on hard soil sites will be more evident, while the same phenomenon exists in soft structures on soft soil sites. Therefore this concept can be utilized in seismic retrofitting of structures by changing structural stiffness so as to reduce seismic action to meet design purposes.

h. To use new seismic technologies. The overall structural seismic behavior can be enhanced by using advanced seismic retrofitting technologies, which should be promoted in seismic retrofitting practice. New research and applications in the United States and Japan have been implemented and there are also some explorations and practices in China during the past decade. The currently available new technologies include base isolation and story isolation, energy dissipation braces and shear walls, and some active control and hybrid control measures. When using the new technologies in seismic retrofitting, the following points should be keep in mind: a) Mature and approved technologies should be used; b) Various comparison and selection should be carried out by relevant professionals through their detailed study and analysis; c) Feasibility in engineering practice to meet site conditions for construction.

CHAPTER 2

Inspection and Evaluation of Building Structures

Generally speaking, the reliability assessments of existing buildings are required before designing a retrofit to provide a technical basis; as well as to avoid design flaws.

2.1 Introduction

2.1.1 Basic Concepts for Reliability Assessment of Buildings

The structural reliability, also called degree of reliability in form of probability, denotes the ability of structures to fulfill the expected functions under specific conditions within a certain service time, which includes the safety, applicability and durability of structures.

The structural safety means the capacity of sustaining different kinds of disasters that may take place under construction and during service time, and of maintaining essential integrated stability when and after accidental events occur. The structural applicability is the capability to realize the scheduled functions under normal service conditions, and the structural durability is the capability to maintain these functions under normal maintenance over time.

Reliability assessment of buildings is to inspect, test and comprehensively analyze the action on the existing buildings, structural resistance and their mutual relationships to assess the practical structural reliability.

2.1.2 Methods of Evaluating Buildings

1. Traditional empiricism

The traditional empiricism used to be employed in reliability assessment of buildings, which is characterized by emphasizing the experience and knowledge of single or a few appraisers, that is, evaluating the structure by personal knowledge and experience following field inspection and necessary checking calculation of experienced technicians. Since no unified standard was available, sometimes the evaluation results depended on the evaluator, especially for complex engineering structures.

2. Practical evaluation method

The evaluation of buildings is being updated and improved with the development of science and technology. The theory of structural reliability has been introduced into the evaluation of building structures, and there has been certain achievement obtained in China. After years of effort, a few standards of reliability assessment for existing buildings were published. The national or industrial standards and specifications already compiled in China for evaluation of existing buildings are listed as follows:

Standard for reliability assessment of industrial factory buildings, GBJ 144-90

Specification for reliability assessment of steel industrial buildings and facilities, YBJ 219-89

Standard for appraisal of reliability of civilian buildings, GB 50292-1999

Technical specification for inspection, assessment and retrofitting of steel structures, YB 9257-96

Standard of dangerous building appraisal, JGJ 125-99

Standard for seismic appraisal of buildings, GB 50023-95
Criteria for the appraisal of seismic performance of industrial structures, GBJ 117-89
Nowadays the practical evaluation method is used in reliability assessment of buildings. It is a scientific evaluation method developed on the basis of the traditional empiricism, in which detailed inspection and analysis are conducted by the inspectors, and the conclusions are given according to current evaluation standards as listed above. Fig. 2.1 provides the working procedures of this method.

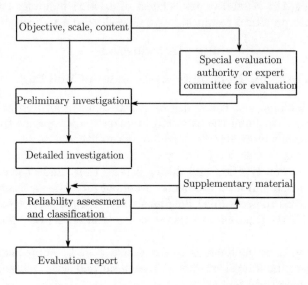

Fig.2.1   Working procedures of practical evaluation method for inspection of buildings.

Presented below is a brief introduction of the work concerned in the procedures of building evaluation using practical evaluation method.

(1) Work of preliminary investigation

During preliminary investigation, the source material, such as the original design and record drawings, should be checked and analyzed for consistence with the built object. Also included is the inspection of the service condition of building structure, which consists of the actions on the structure, service environment and service history.

a. Investigation of action on the structure is to determine the loads and load effects. It can be conducted as indicated in Table 2.1.

Table 2.1   Survey of action on structures

Catalogue	Detailed Item
Permanent load	1. Self weight of structural members, fitting parts and fixed devices 2. Actions of pre-stressing force, soil pressure, water pressure and foundation deformation
Variable load	1. Live load on roof and floor 2. Dust load 3. Hoist load 4. Wind load 5. Snow and ice load 6. Temperature action 7. Vibration impact and other dynamic loads
Accidental load	1. Earthquake 2. Collision and explosion accidents 3. Fire hazard
Other load	Excavation of foundation pit near existing building

The choice of action value can be made in accordance with the stipulation of *Load Code for the Design of Building Structures (LCDBS)*—the current national standard. If an exception occurs or there are no articles concerning the situation in LCDBS, it should be determined in principles prescribed by *Unified Standards for the Design of Structures*.

b. Investigation of service environment of building structures.

Meteorological condition: orientation, wind speed, amount of precipitation, atmospheric humidity and temperature of the building.

Industrial environment: the influence on building structures of liquid corrosion, gaseous corrosion, high temperatures and dampness.

Geographical condition: the influence on building structures of terrain, landform, geologic structure and surrounding buildings.

c. Service history.

Included in this part is usage information related to overload, disaster and corrosion. Special attention should be paid to the load changing history resulting from the variation of service requirements.

(2) Work of detailed investigation

a. Checking for structural layout, brace system, ring beam layout, structural members, structural configuration and joint configuration.

b. Checking for the groundwork and foundation. Excavation or tests may be carried out if necessary.

c. Survey and analysis of action on structure, action effects and effect combination, including statistical field survey if necessary.

d. Inspection and analysis of material performance and geometric parameters of structure, computation and field examination of structural members, including structural test when necessary.

e. Checkout of structural function and configuration of buildings.

The accuracy of inspection has a direct influence on the credibility of structural evaluation, so the inspection of material performance is the basis of reliability assessment, which will be introduced in Section 2.2 of this chapter. For prescription of checkout calculation of concrete, steel and masonry structures or members, refer to Section 2.5.1.

(3) Assessment and classification of reliability

There are two modes for assessment and classification of reliability which will be briefly introduced in Section 2.5.1 and Section 2.5.2 namely:

a. Reliability assessment and classification for building structures.

b. Integrality evaluation and classification for buildings.

3. Evaluation method for buildings in special districts or service environments.

Major differences of existing buildings are:

a. Building age: there are historic buildings built over 1000 years ago, conservation buildings of modern times, and buildings with engineering accidents soon after completion.

b. Location of building: there are seismic regions, areas with collapsible loess, expansive soil and underground excavation areas.

c. Service environment: buildings may be in corrosive environments such as exposure to acidic pollution, carbon dioxide or alkalis; long-term high temperature environment above 100°C or regularly higher than 60°C with heat source; and in vibrating environment.

Given the factors mentioned above, evaluation for buildings located in these special regions or under special service environments should be conducted comprehensively in accordance with industrial and local standards or specifications concerned.

Quite a number of the existing buildings in China, a seismically active country, especially those constructed before 1976, were designed taking no account of earthquake effect. Therefore the evaluation for buildings in earthquake-prone regions should also be in accordance with the standards for seismic appraisal in expectation that the buildings evaluated could maintain a certain margin of safety without collapse, thus the earthquake damage could be relieved and the loss be reduced.

2.2   Inspection for Mechanical Performance of Structural Material

During reliability assessment for buildings it is inevitable to use the mechanical performance parameters of structural material. Although available from the completion record of the buildings, these parameters should be, in most cases, determined through field inspection. The field inspection of existing buildings usually provides basis for solving such problems as: a) Analyzing the cause for damage or failure. b) Assessing the current load-carrying capacity of structure. c) Deciding and optimizing the scheme of treatment and retrofitting for structure. d) Deducing the trend of structural damage development and service life. e) Deducing service life of the structure after retrofitting.

As one of the key parts of reliability assessment, the inspection of structural behavior generally includes examinations of mechanical performance of structural material, structural configuration measurements, size of structural members, position and diameter of reinforcement bars, crack and deformation of structure and members.

The materials used in existing buildings are mainly concrete, steel or rebar, masonry and wood. This section presents an introduction to the inspection methods for the commonly used structural materials such as concrete, steel and masonry.

2.2.1   Concrete Material

Nondestructive inspection technique is generally employed in strength examination for concrete material in existing buildings. Typified by the rebound method and ultrasonic testing method, the nondestructive inspection technique makes use of methods of acoustics, optics, thermal, electromagnetism and radar to measure the physical quantity concerned with the concrete behavior and further to derive its strength and defect without causing damage to its inner structure or service performance. Local damage method for strength inspection of concrete such as core testing and pull-out is also classified as nondestructive, for the result of local damage is confined on the surface of concrete or within a small range of members, having little influence on the integral performance and safety of a structure.

Compared with the conventional destructive tests using standard test blocks, nondestructive inspection is characterized as follows: a) The method is simple and convenient without damaging the members or the structure of building, causing no interference to the serviceability. b) Comprehensive inspection can be conducted directly on the surface of structural concrete in large areas and the quality and strength of the concrete can be reflected more accurately. c) The information not available through destructive tests can be acquired, including inner cavity, loosening, cracking, inhomogeneity, surface burning, freezing damage, chemical corrosion and so on. d) Due to its high applicability, this method is suitable for both new construction and existing buildings. e) The inspection results afford good comparisons with consecutive testing and repeated testing. f) As an indirect testing method, the accuracy of the test results may be relatively low.

Currently in China, five nondestructive inspection methods with corresponding technical standards are in use for testing of the quality and mechanical performance of concrete material. Following is a brief introduction to them.

1. Rebound method for testing compressive strength of concrete

A heavy bob driven by a spring is used to bump the surface of concrete through a knocking lever, and the distance that the bob is bounced back is measured. Then the ratio of bounce distance to initial length of the spring, namely the rebound value, is used as an index to determine the strength of concrete.

The rebound method was first adopted to test the compressive strength of on-site concrete in China in the mid-1950s. The technical symposium on concrete strength testing with rebound apparatus and member testing method was held in 1963, and the book, *Technique for Testing Concrete Strength with Rebound Apparatus*, was published in March 1966, which promoted the rebound method into extensive usage. In the beginning of the 1960s, China started to manufacture the rebound apparatus independently and extend its application, which, however, resulted in errors and misuse because of insufficiency of investigation on various influencing factors and lack of unified technical standard. In 1978 the investigation on nondestructive inspection technique for concrete was listed in the development plan of building science by the National Committee of Construction and a national cooperation research association was formed. Hence, the apparatus performance, influencing factors, testing technique, data processing method and strength deduction method of the rebound apparatus were systematically investigated and the standard state of rebound apparatus and the correlativity among rebound value, carbonization depth and strength were put forward to suit the regional features of China. As a result the test precision and applicability of the rebound method improved. The *Technical Specification for Inspection of Concrete Compressive Strength by Rebound Method* was issued in 1985 and became a professional standard after revision in 1989. The rebound method has become one of the most extensively used methods for nondestructive inspection.

The *Technical Specification for Inspection of Concrete Compressive Strength by Rebound Method* applies to the inspection with moderate-sized rebound apparatus for compressive strength of ordinary concrete in engineering structures. In this technical specification, the related technical requirements, verification and maintenance method are provided. Also included in the specification are the adopted testing technique, methods for measuring rebound value and carbonization depth, computation of rebound values, curves for evaluating strength and calculation for concrete strength. In order to facilitate the proper application of this technical specification by technicians engaged in structural engineering of buildings, the related testing process is explained as follows.

(1) Preparation for Inspection

The concrete structures or members which need inspection in rebound method are usually those without test blocks in same condition or enough standard test blocks, or the quality of the test blocks is not reliable, or test result of the blocks does not meet the requirements of valid technical standard and the result is suspect. Therefore, the inspection should present a comprehensive and correct insight into the tested structure or member.

Before inspection, some information about the structure should be acquired, which includes: the designation of the project, the name of the design unit and construction unit, the physical dimension, quantity and design concrete strength of the structure or member, the type, stability, grade and manufacturer of the cement, the type and grain size of aggregate, the type and quantity of additives, the material measurement, the condition of forming, pouring and curing during construction, the date of set, the reinforcement and prestressing condition, and the service environment and existing problems of structure or members. The most important aspect among these is to find out whether the stability of cement is qualified or not. If not, the structure or member should not be tested by rebound method.

Generally there are two ways to inspect concrete structures or members. One is inspecting structures or members individually, and the other is sampling inspection. Choosing of ways depends on inspection requirements.

The individual inspection mainly applies to independent structures such as cast-in-place shell structures, chimneys, water towers, tunnels and consecutive walls, and individual members such as columns, beams, roof trusses, plates and foundation in which quality and strength of concrete are suspect, and some structures or members are evidently of inferior quality.

The sampling inspection mainly applies to concrete structures or members that are of identical strength grade, basically in the same material, mixing proportion, age, and technical producing conditions. Tested samples should be randomly selected and account for at least 30% of the same kind of structures or members and the total number of test areas should not be less than 100. The detailed sampling method should be specified by agreement of the owner, the construction unit and the inspection unit.

(2) Inspection Method

After learning the condition of concrete structures or members to be tested, the test areas should be chosen and located. The test areas denote every sample to be inspected, which is equivalent to a set of test blocks of the same condition from the sample. Following are some stipulations from the professional standard *Technical Specification for Inspection of Concrete Compressive Strength by Rebound Method*. The concrete from a structure or a member is the minimum element for concrete strength evaluation, and at least 10 test areas should be set. For structures or members shorter than 3 m or lower than 0.6 m, however, the number of test areas can be reduced correspondingly, but under no circumstances be less than 5. The proper size of test area should accommodate 16 rebound test points. The surface of test area should be clean, smooth and dry without seams, finishing coat, stucco layer, laitance, oil crust, voids or pores. Grinding wheels can be used in cleaning work if necessary. Test areas should be evenly distributed on the inspection surface of structures or members without oversize spacing between each other. The spacing between areas could be increased up to a maximum of 2 m if the concrete is of good uniformity and quality. Stress parts and places within structures or members, such as joints of beams and columns, which are prone to generate defects, should be allocated for test. Test areas are preferably located on the sides of concrete perpendicular to casting direction. If unavailable, they should be on the top or bottom surfaces. Test areas should avoid the reinforced bars or embedded pieces set in the vicinity of concrete cover layer. For members with small volume and low stiffness or members thinner than 100 mm at test areas, braces should be installed to support them and prevent them from cracking during rebounding.

With samples selected and test areas laid according to the methods mentioned above, the rebound value is measured first. When testing, the rebound apparatus should be kept perpendicular to the testing surface and should not contact pores or exposed aggregate. Each of the two sides of every test area should be hit with rebound apparatus on 8 test points. In the case that the test area has only one side, it should be hit on 16 points. Every test point receives only one hit with the reading precision of one grade. Test points should be distributed evenly on test area with net space between each other generally not less than 20 mm. The distance between test points and member edges, exposed bars or embedded pieces should be no less than 30 mm.

The carbonization depth should be measured right after rebounding. Proper tool is employed to make a hole into the test area surface with a diameter of 15 mm in which the powder and chipping should be removed (Caution: No liquid can be used to flush the hole.) And immediately, several drips of 1% phenolphthalein alcohol solution are dropped on the inner wall edge of the hole, and the perpendicular distance from the test surface to the edge of the in-hole place where the color does not change into pink is the carbonization depth of this area, which should be measured one to two times, with a precision of 0.5 mm.

Generally one to three places of a test area are selected for carbonization depth measuring. The carbonization depth of one area can be a representative value of its adjacent areas if the concrete quality or the rebound values are basically identical. Commonly no less than 30% of test areas of members are selected for carbonization depth measuring.

(3) Processing of Rebound Data

If the test is horizontal and on cast sides, 3 maximums and 3 minimums in the 16 acquired values of every test area should be ignored and the mean of the 10 remaining values represent the mean rebound value of the test area with one number reserved after decimal point.

Since the curves for strength evaluation in rebound method are acquired through computation of data from horizontally testing sides of concrete members using rebound apparatus, the measured rebound value should be modified if the premises mentioned above cannot be satisfied in testing.

(4) Determination of Concrete Strength from Rebound Data

In general the conversion value of concrete strength responding to its rebound value and carbonization depth at tested area can be found from the conversion table of concrete strength given by the *Technical Specification for Inspection of Concrete Compressive Strength by Rebound Method*. And the putative value of concrete strength can be calculated according to the formula provided by the *Technical Specification for Inspection of Concrete Compressive Strength by Rebound Method*. Attention should be paid that the conversion table in the *Technical Specification for Inspection of Concrete Compressive Strength by Rebound Method* is not applicable to those cases in which the concrete strength is higher than 50 MPa or lower than 10 MPa.

2. Ultrasonic method for inspection of concrete defect

As an important method of non-destructive inspection technique for concrete, the ultrasonic method utilizes the ultrasonic impulse wave–ultrasonic for short hereinafter–to inspect concrete defect, which is based on the fact that some acoustic parameters, such as the propagation time or velocity of the pulse wave and the amplitude and frequency of the receiving wave, when confronting defects, will change relatively in concrete in the same technical conditions (identical raw material, mixing proportion, age and test distance). a) For concrete with certain raw material, mixing proportion, age and test distance, high (or low) velocity of sound indicates high (or low) degree of concrete density because of the direct relationship between them. As a result of the existence of pores and cracks, the integrity of concrete is destroyed and the ultrasonic can only propagate around the pores and cracks to a receiving converter. Therefore the path of propagation lengthens and the propagation time tested increases or the velocity of sound decreases. b) Since the acoustic impedance rate of air is far less than that of concrete, reverberation and scattering occur on the boundary of defect and the acoustic energy decays, especially faster for those of high frequency, when the ultrasonic impulse wave propagating in concrete meets the defects such as voids, pores or cracks. Hence the amplitude, frequency or the high frequency ingredients of received signals decrease markedly. c) The difference of acoustic path and phase exists between the ultrasonic impulse signal reverberated by defects propagating around the direct wave signal. These two kinds of waves interfere with each other after superposition, which results in the distortion of wave pattern of received signal. On the basis of the principle mentioned above, the location and range of concrete defects can be analyzed and distinguished comprehensively, or the size of concrete defects can be estimated using measured values and relative variation of the concrete acoustic parameters.

In the inspection of concrete defects in ultrasonic method the detailed testing procedures are generally determined according to the shape, size and service environment of the structures or members to be tested. Commonly used testing procedures can be classified into several categories as follows.

(1) Testing in plane using thickness vibration energy converter

a. Pair testing: A pair of energy converters, transmitting (T) and receiving (R), is placed respectively on two parallel surfaces of tested structure and the axes of the two–energy converter are collinear.

b. Oblique testing: A pair of energy converters, transmitting and receiving, is placed respectively on two surfaces of tested structure and the axes of the two–energy converter are non-collinear.

c. Single plane testing: A pair of energy converters, transmitting and receiving, is placed on the same surface of tested structure.

(2) Drilling testing using energy radial vibration converter

a. Pair testing in holes: A pair of energy converters is placed respectively at same height in two corresponding holes.

b. Oblique testing in holes: A pair of energy converters is placed respectively at different heights in two corresponding holes and tested with a constant difference in elevation.

c. Plane testing in holes: A pair of energy converters is placed in one hole and tested in synchronized motion with a constant difference in elevation.

The thickness vibration energy converters are often placed on the structural surface to perform a variety of tests, while the redial vibration energy converters are usually placed in drilled holes to execute the pair test and oblique test.

Ultrasonic inspection requires special apparatus and experienced technicians. Available by employing this method are the inspection of distribution and depth of cracks, uncompacted area and cavity, quality of cohesion surface between layers cast in different times and depth of surface damage layer of concrete, uniformity of mass concrete and in certain conditions quality of drilled cast-in-place concrete piles.

The *Technical Specification for Inspection of Concrete Defects by Ultrasonic Method,* issued in 1990 in China, has specified in detail the test apparatus, test technique and test methods for various applications such as inspections of shallow and deep cracks, uncompacted and cavity, quality of concrete cohesion surface, surface damage and uniformity. This has provided assurance for unification of inspection programs, criteria of determination, and improvement of reliability of inspection results.

3. Inspection of concrete strength using ultrasonic-rebound combined method

Two or more kinds of single methods or parameters are combined in inspection of concrete strength. This combined method becomes more and more frequently used in quality control and inspection of concrete because of its smaller error margin and more extensive applicability than single methods. Generally speaking, with the premise of rational selection and combination of single methods, the more nondestructive methods for inspections are used, the more accurate the inspection of concrete strength is.

In the inspection of concrete strength using the ultrasonic-rebound combined method, the ultrasonic and rebound apparatuses are employed to measure the propagation time of sound and rebound value, R, at the same test area of structural concrete, respectively. And then the strength conversion formula already established is utilized to calculate the concrete strength of the test area f_{cu}. Compared with the single ultrasonic or rebound method, the combined method is characterized as follows:

(1) Reducing the influence of age and water ratio

Sound velocity in concrete is affected by factors such as age and water ratio of concrete as well as the coarse aggregate. And apart from the surface state, rebound value is also influenced by age and water ratio of concrete. However, the influence of concrete age and water ratio on sound velocity is essentially different from their influence on rebound value.

For higher water ratio of concrete, the velocity of ultrasonic increases while the rebound value decreases. And for older concrete age, the velocity of ultrasonic increases at a lower rate while the rebound value increases as a result of deeper concrete hardening or carbonization. Therefore the influence of age and water ratio can be reduced partially with the two methods combined in inspection of concrete strength.

(2) Mutually offsetting weaknesses of single method

One physical parameter can reflect concrete mechanics performance only in some aspect to a certain extent, and may lose its sensitivity or may not work beyond this extent. For instance, the rebound value R reflects concrete strength mainly on the basis of the elasticity of surface mortar and this reflection is not sensitive in the case of low concrete strength and large plastic deformation. For members with large sections or great differences of quality in inner and outer parts, it is hardly possible to reflect the real strength. Ultrasonic testing reflects concrete strength according to the dynamic elasticity of whole section. For high strength concrete the elasticity index varies little, so does its corresponding velocity of sound. As a result, slight variation tends to be covered by test error. Thereby the sound velocity V is weakly related to the strength f_{cu}, of tested concrete with strength higher than 35 MPa. When ultrasonic method and rebound method are employed together to inspect concrete strength, the test can be conducted inside and outside and the two methods can mutually offset their weaknesses, when testing concrete of lower or higher strength; thus the real concrete strength can be reflected more comprehensively and authentically.

(3) Enhancing accuracy of inspection

Since it can reduce the influence of some factors and comprehensively reflect quality of whole concrete, the combined method has a remarkable effect on improving accuracy of nondestructive inspection of concrete strength.

Inspection of concrete strength using ultrasonic-rebound combined method was first put forward by the Science Research Institute of Architecture and Building Economy of Romania in 1966. Related technical specifications were compiled, and this method once attracted a lot of attention from scholars of science and technology worldwide. It was introduced into China in 1976 and many research units carried out a considerable number of experiments integrating it in China. Multiple research achievements were obtained in last decade and the method acquired extensive propagation and application in quality inspection of structural concrete engineering. In 1988, the Committee of Standard for Engineering Construction of China approved the first *Technical Specification for Testing Concrete Strength by Ultrasonic-Rebound Combined Method,* which provided important bases for further application of this method.

Technical Specification for Testing Concrete Strength by Ultrasonic-Rebound Combined Method applied to inspecting ordinary concrete compressive strength of building structures using medium rebound apparatus and low-frequency ultrasonic devices. Technical requirements of rebound, ultrasonic apparatus and energy converter, as well as techniques for inspection, operation and maintenance of rebound and ultrasonic apparatus are brought forward at length in this specification. Also stipulated are measurement and computation for rebound value and sound velocity at test areas, and calculation method for strength of tested concrete. The appendix of this specification also lists the fundamental requirements for establishing special or local concrete strength curves, and the conversion table of sound velocity, rebound value and concrete strength at test areas. Like other inspection methods, the ultrasonic-rebound method needs professional technicians to operate devices and make inspection reports while the structural appraiser and retrofitting designer should to a certain extent understand the inspection process and source of data for proper use.

Since inspection of concrete strength with the combined method is essentially an integrated application of the single ultrasonic method and rebound method, the inspection procedures and stipulations concerned are identical to those of the single method mentioned before. For instance, one should prepare materials, understand the tested structures or members in detail and make a detailed plan of disposition of test areas before testing. Testing and computing of tested areas are basically the same as those of single method mentioned before. The concrete strength of tested areas should be calculated or looked up from the tables according to the obtained ultrasonic sound velocity, rebound values and the established relation curves of combined method. The combined method, like other nondestructive inspection methods mentioned before, applies to testing strength of structural concrete under construction as well as concrete in existing buildings.

4. Inspection of concrete strength with drilled core

A special drilling machine is used to bore into structural concrete for sample cores, which will be treated by special devices and then carried through crush test on a machine to examine concrete strength, or observed for quality of inner concrete from external appearance of sample cores. Causing only local damage in structural concrete, drilled core is a quasi-nondestructive field inspection method. Since there is no need for conversion between some physical parameters and concrete strength in inspection using drilled core method, it is generally accepted as a direct, reliable and accurate method. But large-scale sampling tends to be limited to a certain extent because of the inevitable local damage to structural concrete caused during inspection and its high cost. Therefore in recent years it has been suggested worldwide to combine the drilled core method with other nondestructive methods for the purposes of supplementing each other during inspection, that is, nondestructive methods could be employed in large scale, causing no damage to structure on one hand, and the accuracy of nondestructive method could be enhanced by the local-damage drilled core method on the other hand.

Drilled core method is characterized by its visualization and accuracy in inspection of concrete strength and defects such as cracks, seams, delamination, cavities and separation, and is extensively applied to quality inspection of concrete structures or constructions such as industrial or civil buildings, hydraulic dams, bridges, highways and airport runways. In normal concrete structures, cubic test blocks with standard curing process are made for evaluation and acceptance of concrete strength under the requirement of the *Code for Construction and Acceptance of Reinforced Concrete Engineering.* Only in the following circumstances is the inspection of concrete strength by drilling sample cores needed and taken as a primary technical basis for solutions to concrete quality accidents. a) The compressive strength of cubic test block is suspect. b) Quality accidents of concrete structures occur because of inferior quality of cement and aggregate or imperfect construction and curing. c) There is evident difference in quality between surface layer and inner parts at test area of concrete structure, or the members have experienced chemical corrosion, fire and freezing damage during hardening. d) Coarse surfaces of aged concrete structures used for years make it hard to inspect by the rebound or ultrasonic method. e) The thickness of structures or members, such as airport runways and slabs, needs to be inspected.

Although it is characterized by its visualization, reliability, high accuracy and extensive applicability in inspection of concrete strength and quality, drilled core has certain limits as follows: a) The selection of position and number of the drill points are restricted because of the local damage caused during drilling, and the area it represents is also limited. b) Compared with non-destructive inspection apparatus, the drilling machine and the assorted tools for processing sample cores are cumbersome, not portable and the cost of inspection is relatively higher. c) The holes left by drilling for cores need to be filled up and the difficulty in filling increases especially when the rebar is cut off during drilling.

In the *Technical Specification for Testing Concrete Strength with Drilled Core* of the Association of Standardization for Construction Engineering of China, the detailed requirements for the main equipment of drilling and processing sample cores are specified, and the location and method of drilling, number and size of sample cores, the processing and testing for compressive strength of sample cores are also stipulated. Moreover, the calculation formula and conversion table for concrete strength of sample cores with different sizes are provided. The inspection of concrete strength by drilled core method needs professionals to operate and determine concrete strength strictly according to this technical specification. It appears particularly important to test in strict conformity with relevant national technical specifications when analyzing the causes of accidents.

5. Inspection of concrete strength by post-insert pull-out method

The post-insert pull-out method is a micro-damage inspection method conducted after boring, milling, inserting anchor pieces and installing pull-out apparatus on the surface of hardened concrete. With the ultimate pull-out force measured, the concrete strength is determined according to correlativity between pull-out force and concrete strength established in advance. It is characterized by its reliable test results and extensive applicability. The tested concrete strength should be no less than 10 MPa.

The pull-out method can be classified into two categories. One is pre-insert pull-out and the other is pull-out post-insert method. The former is used in field control of concrete quality, such as deciding the proper time for loading or form stripping, which is the most common in construction of concrete cooling towers, as well as the time for applying or releasing prestress, for lifting and transporting members and for stopping moist heat curing or stopping insulation during construction in winter. While the pre-insert pull-out method attained fast popularization and application in north Europe and North America, the post-insert pull-out method is more often employed in China, in which the pull-out test is carried out after boring on hardened concrete and implanting anchor pieces. This method applies to all kinds of members of newly hardened or aged concrete as long as the location of rebar or iron pieces is avoided. In particular it is a highly effective means for inspection when lacking related test information of concrete strength in hand. And it has become one of the inspection methods for field concrete concerned and studied by many countries because of its extensive applicability and high reliability. After more than 10 years of research on the post-insert pull-out method, the *Technical Specification for Inspection of Concrete Strength by Post-insert Pull-out Method,* a national standard, was formulated.

The process of post-insert pull-out method, as illustrated in Fig. 2.2, consists of several steps of boring, milling, installing anchor pieces and pull-out apparatus and pull-out test. The test method depends on the kinds of anchor pieces of pull-out apparatus, as well as parameters such as anchor depth and size of reaction support. The correlativities between pull-out force and concrete strength obtained by different pull-out equipments and operation methods are totally different. The pull-out apparatus currently in use can be generally classified into two large categories. One is the annulus reaction support, which is similar to the LOK and CAPO apparatus manufactured by Denmark, for instance, the TYL pull-out apparatus for concrete strength. The other is the three-point reaction support, developed independently by China. The pull-out equipment of the three-point reaction support, known for easy manufacture, low price and smaller pull-out force than annulus support for concrete of same strength, has a larger inspection range. Like pull-out apparatus of annulus support, it is also a popular device, and both annulus and three-point support devices may be used according to the *Technical Specification for Inspection of Concrete Strength by Post-insert Pull-out Method* of China.

The grain size of coarse aggregate in concrete has the greatest influence on pull-out force in a pull-out test, and the variation coefficient of concrete pull-out force increases with the

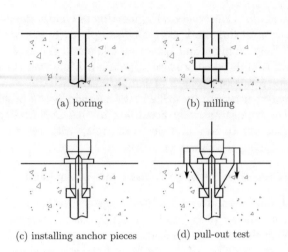

(a) boring   (b) milling

(c) installing anchor pieces   (d) pull-out test

Fig.2.2   Illustrative diagram for post-insert pull-out method.

increasing maximum grain size of coarse aggregate. Therefore it is generally stipulated that the maximum grain size of coarse aggregate in tested concrete should be no greater than 40 mm when anchorage depth of anchoring piece is 25 mm, and for coarse aggregate of greater grain size, deeper anchorage depth is required in order to ensure accuracy of the test results. The grain size of course has influence on the post-insert pull-out test and some anchoring pieces tend to be installed right in the aggregates. Another reason is that different grain sizes result in different volumes of concrete cone pulled out, which is similar to the stipulation for size ratio of coarse aggregate to standard test block. However, the maximum grain size of concrete tends to be no greater than 40 mm in most building engineering in China. Special cases often emerge, especially for those old buildings built before 1949 with great grain size in structural concrete which requires deeper anchorage depth of anchoring pieces. The pull-out test equipment with anchorage depth of 35 mm has been developed in China which meets the requirements of coarse aggregates with maximum grain size no greater than 60 mm and expands the applicability of pull-out test.

In the *Technical Specification for Inspection of Concrete Strength by Post-insert Pull-out Method* of the Association of Standardization for Engineering Construction of China, the detailed technical requirements for pull-out testing equipments are specified, the distribution of test points, testing program and operation steps of pull-out test specified definitely, the specific calculation formula and methods for conversion and extrapolation of concrete strength provided and basic requirements for establishing curves of determining strength by pull-out method also brought forward. The inspection of concrete strength by the post-insert pull-out method, as by other methods, requires special technicians to conduct the test and calibrate the apparatus, and they are expected to have corresponding certificates. Structural engineers should acquaint themselves with the full process of inspection, analyze and make rational use of the test data.

6. Comparison of inspection methods

The rebound method, ultrasonic method, combined method of rebound and ultrasonic, drilled core method and pull-out method are common methods for quality inspection of structural concrete in China. Table 2.2 lists the test content, applicability range, advantages and disadvantages of each inspection method. Selection of these methods can be made comprehensively based on factors such as the existing equipment and field situation.

Table 2.2   Comparison of nondestructive inspection methods

No.	Method	Test contents	Applicability	Characteristics	Disadvantages	Remark
1	Rebound	Surface hardness of concrete	Compressive strength, uniformity of concrete	Simple, quick, without limits in shape and size of tested objects	Test areas confined on concrete surface, repeating test on same place not available	Often applied
2	Ultrasonic	Ultrasonic propagation velocity, amplitude and frequency	Compressive strength, inner defects of concrete	Without limits in shape and size of tested objects, repeating test on same place	Great attenuation of ultrasonic and slightly low accuracy for high probe frequency	Often applied
3	Combined rebound and ultrasonic	Concrete surface hardness and ultrasonic propagation velocity	Compressive strength of concrete	Simple, higher accuracy than single method	Complicated	Often applied
4	Pull-out	pre- or post-install anchoring piece and measure pull-out force	Compressive strength of Concrete	Higher accuracy	Causes certain damage to concrete and needs repair	Often applied
5	Drilled core	Drill samples with certain size from concrete	Compressive strength, splitting strength and inner defects of concrete	Causes certain damage to concrete and needs repair after test	Cumbersome devices, high cost, causes damage to concrete and needs repair	Often applied

2.2.2   Steel or Rebar

In evaluation and retrofitting for existing buildings, steel or rebar needs to be inspected to determine its strength. Particularly when the material performance of rebar is suspected or the rebar performance has changed after the building suffered from disasters or fire, inspection of steel becomes indispensable.

1. Strength inspection of rebar in reinforced concrete structures

Currently mechanical performance of constructional steel or rebar, including ultimate tensile strength, yield strength, extensibility and cold bending behavior, is generally determined in a laboratory by tension test of sample rebar cut from field samples.

(1) Field sampling

Field sampling of rebar is an inspection method causing damage to building structures. Therefore samples taken during field sampling and damage caused to building structures should be as small as possible. Sampling should occur only at unimportant members or subordinate parts of members if possible, for instance, inflection points of beams for tensile rebar. As for alteration and story-adding, sample rebar can be cut directly from parts that are to be removed anyway.

Generally speaking, three samples of each type of rebar should be secured and for suspicious rebar, for instance, the same type of rebar from different batches, the number should be increased according to specific situation.

(2) Inspection of rebar performance

For rebar in reinforced concrete there are requirements for strength, plasticity, cold bending behavior and weldability.

Strength consists of yield strength σ_s, ultimate strength σ_b and yield ratio σ_s/σ_b. Yield strength of rebar is the main basis for calculation in retrofitting design and the yield ratio of rebar may be controlled between 0.60 to 0.75 in order to ensure structures with certain margin of reliability and members with certain ductility.

Plasticity denotes the capability of rebar to generate permanent deformation without fracture under external force, which is expressed by the extensibility δ_5 or δ_{10} at tensile fracture of the sample rebar with gauge length of 5 d or 10 d, respectively.

Cold bending behavior indicates the ability of rebar to bear bending in normal temperature, which is denoted by cold bending angle and diameter of rebar. It is required that no crack, delamination or fracture occur on the convex side of bending area for bending angle between 180° and 90°.

Weldability means that no cracks or oversize deformation take place at rebar after welding of rebar under certain technical conditions, and welding joints maintain performance. Weldability of rebar is influenced by its chemical ingredients.

It is of advantage for the technical performance of rebar if the content of some alloying elements is controlled within certain range. However, excessive content is disadvantageous. For instance, excessive manganese (Mn) and silicon (Si) affect weldability, and excessive carbon (C), titanium (Ti), and vanadium (V) affect plasticity.

Some impurities from raw materials also have influence on rebar performance. Plasticity of rebar decreases with great content of phosphorus (P) and weldability will be influenced by great content of sulfur (S). In addition, the content of oxygen (O) should not exceed 0.05%, nitrogen (N) 0.03% and hydrogen (H) within 0.0003% to 0.0009%.

Mechanical performance and chemical ingredients of rebar in reinforced concrete structures are stipulated in Table 2.3.

Table 2.3   Mechanical performance and chemical ingredients of steel bar

Strength grade	Notation	Nominal diameter (mm)	σ_S (MPa)	σ_b (MPa)	δ_5 (%)	Cold bending d: diameter of bending a: diameter of rebar	C	Si	Mn	V	Ti	P	S	Designation
			No less than									No greater than		
I	R235	8~20	235	370	25	180°, $d = a$	0.14 ~ 0.22	0.12 ~ 0.30	0.30 ~ 0.65	~	~	0.045	0.050	Q235
II	RL335	8~25	335	510	16	180°, $d = 3a$	0.17 ~ 0.25	0.40 ~ 0.80	1.20 ~ 1.60	~	~	0.045	0.045	20MnSi
		28~40	335	490	16	180°, $d = 4a$								
Allowable deviation of chemical ingredients of finished rebar (+)							0.02	0.05	0.10	0.02	0.02	0.005	0.005	

2. Measurement for practical stress of rebar in reinforced concrete structures

Sometimes in investigation and inspection of service state of structures the practical stress of rebar needs to be measured. Introduced below is the measuring method for practical stress of rebar devised by Li Pan.

(1) Method and procedures of measurement

a. Select the maximum stress areas of members as test areas. The practical stresses of rebar in these areas reflect the stress state of the member.

b. Chisel off the cover of tested rebar, shown in Fig. 2.3, and attach strain gauges on the exposed rebar, shown in Fig. 2.4.

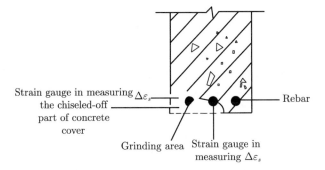

Fig.2.3   Testing practical stress of rebar.

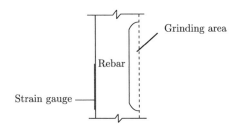

Fig.2.4   Layout of strain gauge.

c. Reduce the rebar section by grinding on the opposite side of the strain gauge. Then measure the reduced amount of rebar diameter with square caliper and at the same time record the increment of strain $\Delta\varepsilon_s$ through strain gauge.

d. The practical stress of rebar σ_s can be calculated according to the formula as follows.

$$\sigma_s = \frac{\Delta\varepsilon_s E_s \bar{A}_s}{\tilde{A}_s} + E_s \frac{\displaystyle\sum_{i=1}^{n} \Delta\varepsilon_{si} A_{si}}{\displaystyle\sum_{i=1}^{n} A_{si}} \leqslant f_y \tag{2.1}$$

where $\Delta\varepsilon_s$ represents the strain increment of ground rebar. $\Delta\varepsilon_{si}$ represents the strain increment of the i th rebar near the tested rebar. E_s represents the elasticity modulus of rebar. $\bar{A}_s$ refers to the section area of ground rebar, shown in Fig. 2.5(a). $\tilde{A}_s$ represents the area being ground off the rebar, shown in Fig. 2.5(b). A_{si} represents the section area of the ith rebar near the tested rebar.

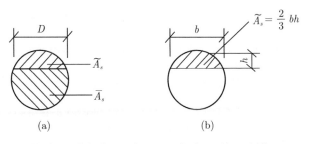

(a)   (b)

Fig.2.5   Calculation for area of rebar after grinding

(2) Test results

Repeat step 3) and step 4) and stop testing until the values of σ_s obtained in two repeats are very close. The present value of σ_s is regarded as practical stress of rebar.

(3) Note

a. The reduced amount of the tested rebar diameter after ground should not exceed one third of the original diameter.

b. The grinding of rebar should be finished in two to four repeats and for every one the reduced amount of rebar section and strain increment at ground area should be recorded.

c. The ground surface should be smooth and the rebar section area after ground should be measured with a square caliper. Record the numerical reading of strain gauge until the temperature of ground surface is identical to that of environment.

d. After the test, weld a rebar of $\phi 20$ with length of 200 mm to the damaged area of ground rebar, then make up the cover with concrete of fine aggregate.

In addition, for a compressive member, with the practical stress σ_s of rebar in it obtained by the method mentioned above, the practical compressive stress σ_c' of concrete on corresponding section could be acquired by

$$\sigma_c' = \left(\frac{E_c}{E_s}\right)\sigma_s \qquad (2.2)$$

where E_c and E_s are the elasticity moduli of concrete and rebar, respectively.

2.2.3   Material for Masonry Structures

Inspection of material strength of masonry structures focuses on mortar, brick or building block of other materials, and integral masonry strength. In comparison with that for concrete material, the technique for material strength inspection of masonry structures is still under development. In recent years, however, considerable research work has been done on it and some standards for inspection are now under the process of approval. And the *Technical Standard for Site Testing of Strength of Masonry Structures* will be issued shortly as a national standard, in which many methods such as the in situ axial compressing method and flat jack are specified for site calculation of compressive strength of mortar, service stress, elastic modulus, compressive and shear strength of masonry. The apparatus of model SQD-1 for testing of mortar strength by point load has been developed by the China Academy of Building Research for supporting the standard. In general the technique for inspection of material strength of masonry structures is approaching maturation day by day. The current inspection methods in common use are briefly introduced as follows.

1. Inspection of integral masonry strength

Integral masonry strength is the main mechanical index of masonry structure. The inspection method for it is as follows.

(1) Site inspection method

Similar to the drilled core method for concrete, the test sample is taken during inspection according to stipulated method directly from field walls–often under sills, and delivered to a laboratory for compressive testing. Since it is difficult to cut samples from walls, and the strength of samples with mortar of low strength is prone to be influenced under even slightest movement, the site testing method is usually replaced by the following methods.

(2) Indirect inspection method

Masonry structure, as a complex body, is composed of mortar and bricks or building blocks of other materials. Therefore its strength may be directly determined on the basis of tested strengths of mortar and brick or building blocks of other materials related to requirements of valid codes.

2. Strength inspection of brick or building block of other materials

The strength of brick or building block of other materials can be determined via samples from masonry structure by conventional method that is relatively simple. The research on brick strength inspection has been conducted by the Building Science Research Institute of Sichuan Province using rebound apparatus of model HT-75 and a specification for strength inspection was also compiled. This method can be used when needed.

3. Strength inspection of mortar

Impact method, point load method and rebound method are often used to inspect the strength of mortar in masonry structures.

(1) Rebound method

Based on the relationship between surface hardness and strength of mortar, the rebound method is nondestructive. Before testing, the plastering, overcoating and cement-mortar filler (for fair-faced walls) on the structure should be removed and small grinding wheels are used to gently smooth the mortar joint. The mortar at test area should have fine adhesion to the bricks and the thickness of mortar joint should be about 10 mm.

In inspection, the rebound apparatus should be aimed at a smoothed mortar joint and perpendicular to the tested side. Then strike five times on the same point consecutively without reading for first two times (known as previous strikes), and get the means of rebound values of the third, fourth and fifth strikes. In the meantime, measure the remaining depth of the little round pit at test area with a precision of 0.1 mm.

The strength of mortar can be determined from some related graphs and tables according to the rebound value N and depth of the pit d. Because of the high discreteness of mortar strength, the rebound test should be carried out on at least five points for each test area and the mean value of these points can be taken as the basis for evaluating the strength.

The rebound method is characterized by its easy operation, quick inspection and portable devices, and it is also a nondestructive inspection method now accepted by inspectors. Its disadvantage is the relatively large deviation of test results. Given these characteristics of rebound method, it is generally thought that more areas should be tested during inspection and a few tests be conducted with other methods such as the impact method. With two methods combined in application, the inspection becomes convenient as well as reliable.

(2) Point load method

A concentrated point load is placed on a mortar layer and the point load carrying capacity of the sample is measured. With the size of sample taken into account, the cubic strength of mortar can be deduced, which makes use of the relationship between the splitting strength and compressive strength of mortar. In inspection, chisel and strike gently for avoiding damage to the sample that may influence its strength.

a. Sample preparation.

Chisel out one mortar layer with two bricks from the structure and take out the mortar layer by striking lightly to make brick drop off or by sawing the brick using manual steel saw or grinding wheel. Then eliminate samples with evident defects and less representativeness. Select samples with uniform thickness and process them into cylinders in diameter of 50 mm (or radius of 15 to 25 mm). At last, smooth the loading or supporting surface carefully.

b. Test procedure.

As shown in Fig. 2.6, both the loading head and support in point load method are round headed cones with radius r of 5 mm. When loading, the upper pressure head should be exactly in alignment with the lower head, and the sample be horizontal. Then increase load slowly until the sample fails. The load P (kN) at failure is read, and the thickness of sample,

t (mm), the distance R (mm) between loading point and sample edges are measured. Thus the strength of mortar, f_m, can be calculated according to the following formula.

$$0.03 f_m^{0.02} + 0.033 = \frac{P}{(0.05R + 1)(0.03t)(0.1t + 1) + 0.4} \tag{2.3}$$

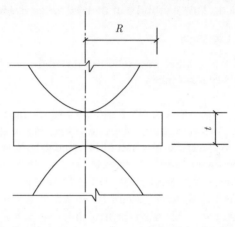

Fig.2.6   Inspection of mortar strength by point load method.

(3) Impact method

During strength inspection of hardened mortar with impact method, impact work is imposed continually on mortar particles, which results in continual crushing and in the process the strength of mortar can be attained.

Main devices in testing include impact apparatus, a round-hole sieve with hole diameters of 12 mm and 10 mm, a set of standard sand sieves and a scale.

When testing, chisel certain amount of mortar from the masonry structure, process it into particles and crush it by impact hammer. The crushing consumes certain amount of energy. After crushed the particles becomes small and fine with surface area increased. Under given impact, the increment of surface area, ΔA, of mortar particles is linearly related to the increment of crushing work, ΔW. And the compressive strength of mortar has a quantitative correlativity with increment of surface area under a unit of work, $\Delta A/\Delta W$, which can be used to determine the mortar strength.

Inspection of mortar strength using impact method is relatively complicated, and when needed, it can be conducted according to appropriate standards.

2.3   Inspection for Reinforcement Disposition in Concrete Members

When doing structural inspection and retrofit design on existing buildings, checking calculations of the strength should be done in terms of actual reinforcement position in structural members, which requires inspection of the location, quantity, diameter and cover thickness of rebar. It is necessary to inspect the actual reinforcement arrangement if there is any suspicion of reinforcement rearrangement, e.g. analyzing the reason why diagonal cracks form in a beam, without low strength of concrete or overstress.

Two types of inspection methods, namely destructive inspection and nondestructive inspection, are commonly used for inspection of disposition, quantity, diameter and cover thickness of rebar.

2.3.1   Destructive Inspection Method (Sampling Inspection)

Generally speaking, the inspection of rebar can be conducted directly on the concrete members. First, chisel off the cover where inspection is needed and directly measure the quantity, diameter and cover thickness of rebar. Then, check the results with the original design drawings. Since this method will cause some damage to the concrete members, it is recommended to chisel off the cover carefully in order to avoid excessive damage to the structure and repair the damage right after the inspection. Generally, the application of this method should be restricted.

2.3.2   Nondestructive Inspection Method

Nondestructive inspection methods, which do not influence the inner structure and performance of concrete, determine the disposition of rebar and cover thickness using sound, light, heat, electricity, magnetism and radiation. With the development of inspection techniques, nondestructive inspection methods have reached a new level. So far, electromagnetic, radar and ultrasonic methods. are mainly used as nondestructive inspection methods at home and abroad.

1. Introduction of each method

Electromagnetic testing is conducted according to optical principles, which consists of electromagnetic wave (microwave) and electromagnetic induction methods. Electromagnetic induction can be used to inspect magnetic substance such as rebar. Inspection of the location of rebar and cover thickness is mostly conducted in structural assessment and retrofit design, and the most common apparatus is a rebar detecting instrument.

Rebar detecting uses electromagnetic induction, which is based on the principle of the decreasing of the voltage amplitude for a parallel resonance circuit. The principle of electromagnetic induction is shown in Fig. 2.7. Coils are set in the detecting device, such as in a probe, in which a magnetic field will be generated when alternating current goes through. If there are some magnetic substances such as rebar in the magnetic field, current will be generated in these magnetic substances, with which an inverse magnetic field will be consequently generated. Because of the new magnetic field, inverse current will be generated in the coil, which will cause the variation of coil voltage. Since the coil voltage changes in accordance with the variation of the characteristics of the magnetic substance (rebar) and the spacing between, the location of rebar and cover thickness can be determined by checking these changes. The instrument of rebar detector is shown in Fig. 2.8.

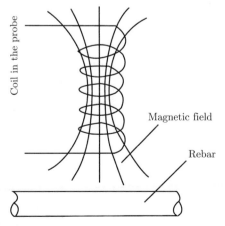

Fig.2.7   Principle of the electromagnetic induction method.

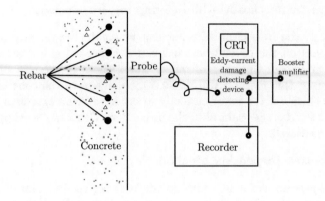

Fig.2.8   Schematic of rebar detecting instrument.

Currently, there are various kinds of rebar detectors such as PROFOMETER 4 Rebar Locator (Switzerland). The made-in-China rebar detectors are shown in Table 2.4. The GBH-1 rebar detector is shown in Fig. 2.9.

The principle of the radar instruments for detecting the reinforcement disposition in concrete is as follows: the electromagnetic wave which is transmitted from the radar antenna is reflected from the interface of the substance, such as rebar, which has different electric characteristics from concrete, and then return to the antenna located on the concrete surface.

Table 2.4   Detectors of the location and cover thickness of rebar

Name and model	Weight & dimension	Application	Display mode	Technique parameters	Manufacturer
HBY-84A detector of concrete cover thickness	250 mm× 120 mm× 100 mm	Inspecting location and cover thickness of rebar	Dial	Measuring range: 5 ∼ 60 mm, detective error when cover thickness less than 45 mm: ±3 mm	Shandong Sanlian Electronic Corporation
GBH-1 detector of concrete cover thickness	220 mm× 120 mm× 120 mm, weighing 2.5 kg	Inspecting position and cover thickness of rebar	Dial	Measuring range: 12 ∼ 120 mm, detective error when cover thickness is 20 ∼ 80 mm: ±3 mm	Shantou Electronic Devices Factory; Ultrasonic Electronic Devices Corporation
GB-1 smart detector for location and concrete cover thickness of rebar	340 mm× 120 mm× 280 mm, weighing 2.5 kg	Inspecting location and cover thickness of rebar	Digital	Measuring range 0 ∼ 68 mm, error: 0 ∼ 30: ±1 mm; 30 ∼ 50: ±2 mm; 50 ∼ 68: ±3 mm	Tongji University (Shanghai Institute of Building Materials)
GBY-1 detector of cover thickness	220 mm× 150 mm× 100, weighing 3.25 kg	Inspecting location, diameter and cover thickness of rebar	Digital	Measuring range of rebar diameter: $\phi 6 \sim 50$ mm measuring range of cover thickness: 0 ∼ 170 mm, error: 0 ∼ 60: ±1 mm; 60 ∼ 170: ±3 mm; 120 ∼ 170 : ⩽ 10%	Highway Scientific Research Institute of Ministry of Communications, Beijing Qingyun Automation Technology Development Corporation

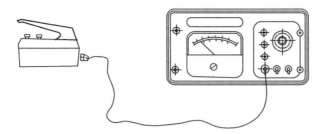

Fig.2.9   GBH-1 rebar detector.

According to the interval time between transmitting and returning of the electromagnetic wave, the distance between the reflecting body and the concrete surface can be determined, which means the location and cover thickness of rebar can be detected. Radar can display the image of the rebar in the concrete section consecutively along the measuring line. This procedure can be conducted in a short time.

The ultrasonic method, used as nondestructive inspection, can acquire the interior information of the inspected object through the medium of ultrasound. The principle of this method is as follows: contact the transmitting probe (electric-acoustic transducer) and receiving probe (acoustic-electric transducer), which consists of piezoelectric elements, to the concrete surface; then the transmitting probe transmits ultrasonic wave, and the receiving probe will receive it; the location of rebar and cover thickness can be detected according to the acoustic parameters of the receiving ultrasonic wave.

2. Applicability and comparison of each method

Table 2.5 shows the methods and instruments of inspection of the location of rebar and cover thickness by using nondestructive techniques.

If it can be concluded from Table 2.5 that different nondestructive inspection methods (electromagnetic method or radar method) can be used in the same inspection item. Which method is more effective and more convenient? Generally speaking, each method has its own advantages in different cases due to various influencing factors. According to the survey of the results using the methods mentioned above, some conclusions could be drawn as below:

Table 2.5   Nondestructive inspection method and detecting devices

Inspection Item	Method	Device	Remarks
Rebar location	Electromagnetic	Rebar detector	General domestic manufacturers
	Radar	Electromagnetic-wave inner concrete detector	
Cover thickness	Electromagnetic	Rebar detector	General domestic manufacturers
	Radar	Electromagnetic-wave inner concrete detector	
	Ultrasonic	Ultrasonic detector	

a. Radar is quick, and electromagnetic is slower.

b. There is no big difference in precision when detecting the rebar position. Both of them are applicable.

c. Ultrasonic method has relatively higher precision for detecting the cover thickness, and it will have greater error when electromagnetic method and radar method are used to detect the rebar of smaller diameter and thicker cover.

d. During the procedure of inspecting rebar in concrete, it is suitable to determine the rebar location with electromagnetic method and radar method first, and then detect the cover thickness with ultrasonic method.

The methods mentioned above cannot accurately detect the diameter or connection of rebar in the joint and members. However, these detecting results are the main bases for structural evaluation and retrofit design. With the increasing projects of structural evaluation and retrofitting design, it is urgent to develop high-precision inspection instruments. Though some domestic research units have made some progress on the detective technique of rebar imbedded in joints, these techniques have not reached the level of practical utilization.

2.4   Deformation Inspection for Structures and Members

In structure evaluation and rectification for existing buildings, deformation (deflection) inspections of some structure members, as well as inclination and settlement (rate of settlement) measurements, are always necessary.

2.4.1   Deformation Measurement of Structure Members

Deformation measurement of structure members is indispensable for the inspection of existing buildings, especially for the beams and slabs that have quality problems or long histories. The deformation measurement of the structure members introduced here is mainly about the deflection measurement of the beams and slabs, which includs the following methods.

(1) In situ static load test

Static load test (nondestructive test) is conducted on the beams and slabs, and then the deflection of the beams and slabs during the loading process is measured. There are two loading modes, piling the sands, stones, bricks, stone blocks or other clogs on the beam or slab to form a uniform load, and putting the clogs on a loading plate and hanging it on the beam with a suspender to form a concentrated load.

The water-loading mode is another option. During the water loading process, load piling, unloading or load weighing is not needed. Thus the amount of work is reduced. However, a trough is required to prevent water leakage. Waterproof materials like plastic films should be used as the inner lining. The size of the trough should be determined by the weight of water that required. Generally, the depth of the trough should be about 200 mm greater than the depth of the required water.

The hydraulic jack can also be used in loading; the hydraulic loading method is widely used in load test, and it can be used to apply different kinds of loads, such as uniform load, concentrated load, and unsymmetrical load. But the reaction force produced by the hydraulic jack must be balanced.

Fig. 2.10 and Fig. 2.11 show the piling loading and water loading modes.

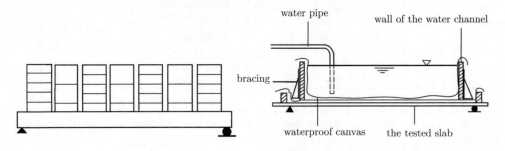

Fig.2.10   Piling loading mode.   Fig. 2.11   Water loading mode.

Dial indicator or displacement meter is always used to measure the deflection, and the dial indicator is the mechanical displacement meter. At present, electric displacement meter used in static load tests has advantages such as it has a wide range, allows reading and measuring from a long distance, recording automatically, and directly transmitting the result signal into the computer to perform the data collecting and processing, all of which make the measurement convenient.

As for the arrangement of test points, besides those in the areas with the maximum deflection, test points should also be set at each end (support) of the members to measure the deformation, and the corresponding errors should be deducted when analyzing the measured data.

(2) Using level gauge

The method of measuring the deformation of the beams and slabs with a level is as follows: Set the surveyor's poles vertically at the supports and mid-span of the beams and slabs, measure the reading on the leveling poles at the same height, and compare the reading of the support and the mid-span to get the deflection of mid-span of the member. As there might be some errors in the process of setting the leveling poles and measuring, it is difficult to achieve a precise deflection value.

(3) Another way to measure the deflection of the mid-span of beams and slabs

Tighten a steel wire or a chord wire between the supports of beams or slabs, then measure the distance from the wire to the member surface in the mid-span to get the deflection of the mid-span of the member. Generally speaking, tightening will directly affect the measuring result; this method will cause a bigger error.

2.4.2   Inclination Inspection of Buildings

The outer corners of the buildings can be considered as the observation points for inclination inspection. Generally, the inclination inspection should be conducted on all four outer corners of the building. After a comprehensive analysis, the inclination of the whole building can be determined. Now, theodolite is the most widely used device for inclination inspection.

(1) Determination of the position of theodolite

The position of theodolite is shown in Fig. 2.12; the distance between the theodolite and the building (L) should be larger than the building height.

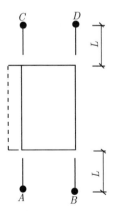

Fig.2.12   Inclination inspection of buildings.
(The continuous lines in the figure indicate the original building, while the dashed lines indicate the inclined building.)

(2) Measurement of the inclination

As shown in Fig. 2.13, aim at the point M at the top of the wall corner, cast down to the point N, measure the horizontal distance A of NN', then take point M as a reference point, and get the angle α with theodolite.

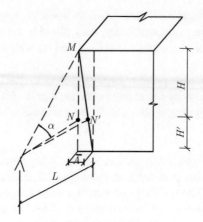

Fig. 2.13   Measuring method.

(3) Analysis and computation of the inclination data

According to the vertical angle α, height can be calculated by the following equation:

$$H = L \cdot \text{tg}\alpha \qquad (2.4)$$

Then the inclination of the building will be:

$$i = A/H \qquad (2.5)$$

The inclination of the outer corner is:

$$\overline{A} = i \cdot (H + H') \qquad (2.6)$$

By repeating steps (2) and (3), the gradients of all four outer corners can be obtained, which can comprehensively reflect the inclination of the whole building. During rectification for existing buildings, it is necessary to do the inclination inspection.

2.4.3   Settlement Inspection of Buildings

When doing structural retrofitting design, it is necessary to learn about the settlement of a building, including the rate of settlement and nonuniform settlement. If the original information about the settlement is available, the accumulated settlement could be acquired. For the buildings with unsteady rate of settlement, the superstructure cannot be retrofitted before the foundation is strengthened. The settlement inspection is very important for retrofitting design.

Level is the most widely used device for settlement inspection. Today, the optical sensor is available, which means the optical sensor technology has been applied to the settlement inspection. In this section, the method of settlement inspection with a level will be mainly discussed.

In order to assure the measuring precision, it is suitable to employ the level II. Do not change the measuring tools or the surveyor during the survey process. In order to obtain continuous settlement data, do not change the reference point and the elevation arbitrarily, and calibrate the devices accurately.

(1) Arrangement of reference points

Use the reference points of the existing building to determine the accumulated settlement after measurement. The principle of arranging the reference points is to ensure the stability. Two or three exclusive reference points for settlement inspection need to be embedded at the proper positions near the building. To prevent effects from the construction of the building and the base pressure, the reference points should not be set too close to the building. To eliminate the effect of different elevation of the ground caused by settlement, the reference points should not be too far from the building, usually not over 100 m.

(2) Arrangement of the observation points

The numbers and positions of the observation points should reflect the history of the building settlement in an all-round way, while factors as the building shape, structure, engineering geological conditions and settlement modes should be comprehensively considered, and the position should be easy to observe from and preserve. At least six observation points are usually set with spacing of 15~30 m along the building. Furthermore, observation points should also be arranged at positions where foundation patterns or geological condition changes or suffers heavy load. When observing the existing buildings, the former observation positions can be used; if the nonuniform settlement had existed, the observation position could be arranged according to the situations on site, and generally be set where the largest settlement is formed. The settlement observation point made of thick rebar is usually set on the walls of the building, as shown in Fig. 2.14.

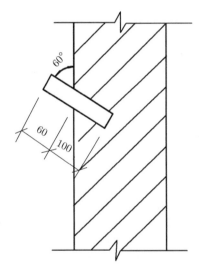

Fig. 2.14   Observation points of settlement.

(3) Settlement data analysis

Use level and leveling staff to measure and read the elevation of each observation point, calculate the elevation of each settlement point immediately after settlement observation, and acquire the present settlement, accumulated settlement and the rate of settlement. According to the settlement of each observation point, the differential settlement of each point will be obtained. As a result, nonuniform settlement data of the building will be acquired. If necessary, the relation curve of load (P), settlement (S), time (T) as well as the settlement distance (L) relation curve, can be plotted according to the observation data at each stage (shown in Fig. 2.15), for evaluation and retrofitting design of the existing building.

When dealing with an engineering accident, the present nonuniform settlement of the building needs to be measured. Since nonuniform settlement is obvious, the earth covering the surface of the foundation at the position of the largest settlement can be moved away,

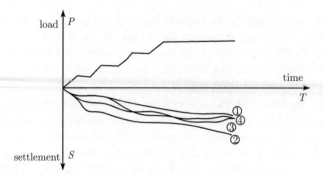

Fig. 2.15   Curve of load-settlement time.

and the observation point can be set at the surface of the foundation. When measuring, the level will be arranged at the place with same distances to the two observation points, then place the leveling staff at the position of the points (on the surface of the foundation), obtain the reading of the elevation at the same level, then the nonuniform settlement will be acquired. With the same method, nonuniform settlement between any two observation points can be acquired, and the present nonuniform settlement of the whole building will be known.

2.4.4   Inspection of Masonry Cracks

There are many reasons for the occurrence of masonry cracks, such as settlement cracks, temperature cracks, load cracks and cracks caused by natural disasters like fire and earthquake. These cracks have great influence on the bearing capacity, service performance and durability of the masonry structure. Thus, the cracks should be inspected in an all-round way. The inspection includes the position, amount, width, length, direction, pattern, and stability of the crack.

It is simple to inspect the crack length. The work can be done using measuring tools like ruler and steel tape. The crack width can be measured by the crack-width comparison card, calibration magnifier, and special crack width-measuring instrument.

The positions, numbers, directions and patterns of the cracks can be inspected visually, and marked on the elevation drawings of the wall and photographed.

The active cracks should be inspected regularly, and sticking plaster bandage on the cracks is the simplest and most widely used way in current inspection. The stability of the cracks can be determined by observing the cracks of the plaster bandage. If it is an active crack, the largest width and length of the cracks at different times should be recorded at the positions of the cracks. The change of the length could be recorded by regularly marking at the end of each crack.

2.5   Structural Reliability Assessment

At present, there are two practical methods of reliability assessment: building structure reliability assessment method and building damage degree assessment. A simple introduction of the two methods is made in the following sections.

2.5.1   Building Structure Reliability Assessment Method

The structural reliability levels are determined by the state of building structure reliability (safety, applicability, durability), which is defined as the "three levels and four grades" assessment method.

This method is based on the reliability theory. With rigorous and rational theoretical foundation, it is advanced in theory and called the *Industrial Factory Building Reliability Assessment*. Though this assessment method does not achieve the dependability of approximate probability method (level II) or include the analysis of numerical calculation of structural reliability, it is relatively convenient and applicable.

1. The standards of assessment

Building structure reliability assessment can be divided into three levels, namely sub-item, item or combined item, and unit, while each level can be divided into four grades as shown in Table 2.6.

Table 2.6   Grades and levels of building structure reliability assessment

Level	Unit	Item or combined item			Sub-item
Grades	1, 2, 3, 4	A, B, C, D			a, b, c, d
Scope and contents	Unit	Structural layout and bracing system		Structural layout and brace layout	
				Slenderness ratio of bracing system	Slenderness ratio of bracing member
		Load-bearing system	Ground & foundation	Ground, slope	
				Foundation	According to corresponding structure
				Piles and pile foundation	Piles and pile foundation
			Concrete structure		Bearing capability, structural details, connections, cracks, and deformation
			Steel structure		Bearing capability, structural details, connections, deformation, and deviation
			Masonry structure		Bearing capability, structural details, connections, deformation, cracks, and deformation
		Enclosure structure system	Service function		Roof system, walls, doors and windows, underground waterproof facilities, enclosure facilities
			Bearing structure		According to corresponding structure

Relevant provisions and methods in the standard of the assessment are introduced in the following paragraphs.

(1) Concept of sub-item, item and unit

a. Sub-item.

Sub-item is the first level of building reliability assessment. It means ground, foundation, pile and pile foundation, and slopes; for concrete structure, steel structure, masonry structures or members, it refers to bearing capability, structural details, connections, deformation and cracks; for enclosure structure systems, according to service function, sub-items include: roof system, walls, doors and windows, underground waterproof facilities, and protective facilities.

Table 2.7   Grading of slenderness ratio of steel brace

Types of factory buildings	Types of brace members		Slenderness ratio of brace members			
			a	b	c	d
With medium or light cranes, or without crane	Ordinary brace	Tension members	$\leqslant 400$	$>400, \leqslant 425$	$>425, \leqslant 450$	>450
		Compression members	$\leqslant 200$	$>200, \leqslant 225$	$>225, \leqslant 250$	>250
	Lower column bracing	Tension members	$\leqslant 300$	$>300, \leqslant 325$	$>325, \leqslant 350$	>350
		Compression members	$\leqslant 150$	$>150, \leqslant 200$	$>200, \leqslant 250$	>250
With heavy crane or with hammers rating $\geqslant 5$ t	Ordinary brace	Tension members	$\leqslant 350$	$>350, \leqslant 375$	$>375, \leqslant 400$	>400
		Compression members	$\leqslant 200$	$>200, \leqslant 225$	$>225, \leqslant 250$	>250
	Lower column bracing	Tension members	$\leqslant 200$	$>200, \leqslant 225$	$>225, \leqslant 250$	>250
		Compression members	$\leqslant 150$	$>150, \leqslant 175$	$>175, \leqslant 200$	>200

Notes: 1) The ordinary brace listed in the table refers to all braces except lower column bracing.

2) For the bracing systems resisting dynamic load directly or indirectly, the least radius of gyration of the angle steel should be used in calculation of the slenderness ratio of the tension member. But the axis of the radius of gyration should be parallel to the leg of angle steel in calculation of the out-of-plane slenderness ratio of the cross tension member.

3) In a factory building with crab crane or rigid crow crane, the slenderness ratio of bracing tension members should generally be assessed according to the slenderness of the tension member of the lower column bracing in the factory "with medium or light duty cranes, or no crane" in the table.

4) For the factory building bearing greater dynamic load, the assessment of slenderness ratio of brace members should be stricter.

5) With adequate experience, the assessment of slenderness ratio of lower column bracing in the factory building can be relaxed.

6) When the slenderness ratio of compression member of lower inter-column cross brace is comparatively great, it is feasible to check the computations according to the tension member and make assessment according to the slenderness ratio of tension members.

Since the grading of each sub-item is based on the limit state of certain function, assessment of sub-items is based on whether structural members satisfy the single functional requirement (reliability requirement), which is expressed as a, b, c, or d. In addition, according to impact on reliability of item, sub-items can be divided into principal sub-items and secondary sub-items.

Sub-items (bearing capability, deformation of structures or members, slenderness ratio of steel brace, etc.) can be graded according to Table 2.7–Table 2.18.

b. Item or combined item.

Item or combined item is the second level in structural reliability assessment, which can be divided into basic item and combined item according to composition. Ground foundation, structure and structure members are basic items; load-bearing system, structural layout, bracing system and enclosure structure system are combined items. All the items, except structural layout and bracing system that have no sub-items are evaluated by the second level assessment, according to the evaluated results of sub-items. Therefore, the assessment is the integrated evaluated result of whether structure member or structure system satisfies each functional requirement, which will be expressed as A, B, C, or D.

Table 2.8   Assessment of bearing capability of structures or members

No.	Types of Structures or members	Bearing capability $R/\gamma_0 S$			
		a	b	c	d
1	Reinforced concrete roof trusses, brackets, roof beams, platform girders and columns, medium or heavy crane beams; general members and braces of steel structure; masonry structures or members	$\geqslant 1.0$	$<1.0, \geqslant 0.92$	$<0.92, \geqslant 0.87$	<0.87
2	Steel roof trusses, joists, beams, columns, connections and details of medium or heavy crane beams	$\geqslant 1.0$	$<1.0, \geqslant 0.95$	$<0.95, \geqslant 0.90$	<0.90
3	General reinforced concrete members (including floors, cast-in slabs, beams, etc.)	$\geqslant 1.0$	$<1.0, \geqslant 0.90$	$<0.90, \geqslant 0.85$	<0.85

Notes: 1) If steel member or connector has fractures or sharp angle notches, it will be rated as c or d according to the basic principle of assessment, based on its impact on bearing capacity.

2) For welding crane beam, if fatigue cracking appears in or near the connective weld at top flange, or appears in the transverse welding seams of the web in tensile region at the ends of stiffening ribs or the tensile flange; or if there are any steel components welded to the tensile flange, it will be rated as c or d.

3) For masonry structures or members that have obvious stress cracks in compression, in bending, in shear, etc., it will be rated as c, or d, according to the basic principle of assessment based on degree of damage.

Table 2.9   Assessment of crack width of concrete structures or members reinforced with rebar of Grade I, II, III

No.	Service conditions of structures or members		Crack width (mm)			
			a	b	c	d
1	Normal indoor environment	General members	$\leqslant 0.40$	$>0.40, \leqslant 0.45$	$>0.45, \leqslant 0.70$	>0.70
		Roof trusses and brackets	$\leqslant 0.20$	$>0.20, \leqslant 0.30$	$>0.30, \leqslant 0.50$	>0.50
		Crane beams	$\leqslant 0.30$	$>0.30, \leqslant 0.35$	$>0.35, \leqslant 0.50$	>0.50
2	Outdoor or high humidity indoor environment		$\leqslant 0.20$	$>0.20, \leqslant 0.30$	$>0.30, \leqslant 0.40$	>0.40

Note: "Outdoor or high humidity indoor environment" refers to the structures or members exposed to following service conditions: rain, always affected by steam and condensate water indoors, or contacted with soil directly.

Table 2.10   Assessment of crack width of prestressed concrete structures or members reinforced with rebar of Grade II, III, IV

No.	Service conditions of structures or members		Crack width (mm)			
			a	b	c	d
1	Normal indoor environment	General members	$\leqslant 0.20$	$>0.20, \leqslant 0.35$	$>0.35, \leqslant 0.50$	>0.50
		Roof trusses and brackets	$\leqslant 0.05$	$>0.05, \leqslant 0.10$	$>0.10, \leqslant 0.30$	>0.30
		Crane beams	$\leqslant 0.05$	$>0.05, \leqslant 0.10$	$>0.10, \leqslant 0.30$	>0.30
2	Outdoor or high humidity indoor environment		$\leqslant 0.02$	$>0.02, \leqslant 0.05$	$>0.05, \leqslant 0.20$	>0.20

Table 2.11   Assessment of crack width of prestressed concrete structures or members reinforced with carbon steel wire, steel wire, heat-treated rebar, cold-drawn low carbon wire

No.	Service conditions of structures or members		Crack width (mm)			
			a	b	c	d
1	Normal indoor environment	General members	$\leqslant 0.02$	$>0.02, \leqslant 0.10$	$>0.10, \leqslant 0.20$	>0.20
		Roof trusses and brackets	$\leqslant 0.02$	$>0.02, \leqslant 0.05$	$>0.05, \leqslant 0.20$	>0.20
		Crane beams	-	$\leqslant 0.05$	$>0.05, \leqslant 0.20$	>0.20
2	Outdoor or high humidity indoor environment		-	$\leqslant 0.02$	$>0.02, \leqslant 0.10$	>0.10

Table 2.12   Assessment of deformation of concrete structures or members

Types of Structures or members		Deformation			
		a	b	c	d
Roof trusses and brackets (single story factory)		$\leqslant L_0/500$	$> L_0/500,$ $\leqslant L_0/450$	$> L_0/450,$ $\leqslant L_0/400$	$> L_0/400$
Multistory frame girders		$\leqslant L_0/400$	$> L_0/400,$ $\leqslant L_0/350$	$> L_0/350,$ $\leqslant L_0/250$	$> L_0/250$
Other roofs, floors and stair members	$L_0 > 9$ m	$\leqslant L_0/300$	$> L_0/300,$ $\leqslant L_0/250$	$> L_0/250,$ $\leqslant L_0/200$	$> L_0/200$
	7 m $\leqslant L_0 \leqslant 9$ m	$\leqslant L_0/250$	$> L_0/250,$ $\leqslant L_0/200$	$> L_0/200,$ $\leqslant L_0/175$	$> L_0/175$
	$L_0 < 7$ m	$\leqslant L_0/200$	$> L_0/200,$ $\leqslant L_0/175$	$> L_0/175,$ $\leqslant L_0/125$	$> L_0/125$
Crane beams	Electric crane	$\leqslant L_0/600$	$> L_0/600,$ $\leqslant L_0/500$	$> L_0/500,$ $\leqslant L_0/400$	$> L_0/400$
	Manual crane	$\leqslant L_0/500$	$> L_0/500,$ $\leqslant L_0/450$	$> L_0/450,$ $\leqslant L_0/350$	$> L_0/350$
Multistory factory under wind load	Horizontal interstory deformation	$\leqslant h/400$	$> h/400, \leqslant h/350$	$> h/350, \leqslant h/300$	$> h/300$
	Total horizontal deformation	$\leqslant H/500$	$> H/500, \leqslant H/450$	$> H/450, \leqslant H/400$	$> H/400$
Out-of-plane inclination of bent columns of single story factory		$\leqslant H/1000,$ and $\leqslant 20$ mm when $H > 10$ m	$> H/1000,$ $\leqslant H/750,$ and > 20 mm, $\leqslant 30$ mm when $H > 10$ m	$> H/750,$ $\leqslant H/500,$ and > 30 mm, $\leqslant 40$ mm when $H > 10$ m	$> H/500,$ and > 40 mm when $H > 10$ m

Notes: 1) L_0 is the effective span of member. H is the total height of column or frame. h is the story height of frame.

2) The deformation values listed in this table result from the long-time load effect combination, which should decrease or increase the value of fabricating invert arch or down deflection.

Table 2.13   Assessment of deformation of steel structures or members

Types of steel structures or members		Deformation			
		a	b	c	d
Purlins	Light roofs	$\leqslant L/150$	>level "a" deformation, having no functional influence	>level "a" deformation, having partial functional influence	>level "a" deformation, having functional influence
	Other roofs	$\leqslant L/200$			
Trusses, roof trusses and brackets		$\leqslant L/400$	>level "a" deformation, having no functional influence	>level "a" deformation, having partial functional influence	>level "a" deformation, having functional influence

Continued

Types of steel structures or members		Deformation			
		a	b	c	d
Solid beams	Girders	$\leqslant L/400$	>level "a" deformation, having no functional influence	>level "a" deformation, having partial functional influence	>level "a" deformation, having functional influence
	Other beams	$\leqslant L/250$			
Crane beams	Light and medium duty $(Q < 50\ t)$ bridge cranes	$\leqslant L/600$	>level "a" deformation, having no functional influence on service of crane	>level "a" deformation, having partial functional influence on service of crane, can be remedied	>level "a" deformation, having functional influence on service of crane, cannot be remedied
	Heavy and medium $(Q > 50\ t)$ bridge duty	$\leqslant L/750$			
Columns	Transverse deformation of factory columns	$\leqslant H_T/1250$	>level "a" deformation, having no functional influence on service of crane	>level "a" deformation, having partial functional influence on service of crane	>level "a" deformation, having functional influence on service of crane, cannot be remedied
	Transverse deformation of outdoor trestle columns	$\leqslant H_T/2500$			
	Longitudinal deformation of factory and trestle columns	$\leqslant H_T/4000$			
Wall-frame members	Cross beams sustaining masonry (horizontal)	$\leqslant L/300$	>level "a" deformation, having no functional impact	>level "a" deformation, having functional impact	>level "a" deformation, having severe functional impact
	Light wall cross beam (horizontal) as profiled steel sheets, corrugated iron	$\leqslant L/200$			
	Post shorings	$\leqslant L/400$			

Notes: 1) "L" in this table is the span of bending member. "H_T" is the height from bottom surface of column to top surface of crane beam or crane truss. Deformation of column is the horizontal deformation resulting from horizontal load of the heaviest crane.

2) The deformation values listed in this table result from the long-time load effect combination, which need minus or plus value of fabricating invert arch or down deflection.

Table 2.14   Assessment of crack width of masonry structures or members

No.	Structures or members	Deformation crack			
		a	b	c	d
1	Walls and pilastered walls	No crack	With slight cracks on wall, widest crack width<1.5 mm	With considerable cracks on wall, widest crack width between 1.5 mm and 10 mm	With severe cracks on the wall, the widest crack width>10 mm
2	Independent columns	No crack	No crack	Widest crack width≤1.5 mm, and not through column cross section	With column fracture or horizontal displacement

Note: This table is only applicable to masonry structures made of clay bricks, silica bricks, and fly ash bricks.

Table 2.15   Assessment of deformation of single story factory of masonry structures or members

No.	Member types	Deformation or inclination value Δ(mm)			
		a	b	c	d
1	Walls and columns of factories without cranes	$\leqslant 10$	$> 10, \leqslant 30$	$>30, \leqslant 60$ or $\leqslant H/150$	> 60 or $> H/150$
2	Walls and columns of factories with cranes	$\leqslant H_T/1250$	Having inclination, but having no influence on function	Having inclination, and having influence on service of crane, but adjustable	Having inclination, and having influence on service of crane, not adjustable
3	Independent columns	$\leqslant 10$	$> 10, \leqslant 15$	$> 15, \leqslant 40$ or $\leqslant H/170$	> 40 or $> H/170$

Notes: 1) In this table "H_T" is the height from bottom surface of column to top surface of crane beam or crane truss; Δ is the deformation or inclination value of masonry wall or column of single story factory; "H" is the total height of the masonry building.

2) This table is applicable to the situations that the height of wall or column $\leqslant 10$ m. When the height of wall or column>10 m, the acceptable deformation or inclination value of each level can increase 10% with every 1 m accretion in H.

Table 2.16   Assessment of deformation of multistory factory of masonry structures or members

Member types	Inter-story deformation or inclination δ(mm)				Total deformation or inclination (mm)			
	a	b	c	d	a	b	c	d
Walls and pilastered walls	$\leqslant 5$	$> 5, \leqslant 20$	$> 20, \leqslant 40$ or $\leqslant h/100$	>40 or $> h/100$	$\leqslant 10$	$> 10, \leqslant 30$	$> 30, \leqslant 60$ or $\leqslant H/120$	> 60 or $> H/120$
Independent columns	$\leqslant 5$	$> 5, \leqslant 15$	$> 15, \leqslant 30$ or $\leqslant h/120$	>30 or $> h/120$	$\leqslant 10$	$> 10, \leqslant 20$	$> 20, \leqslant 45$ or $\leqslant H/150$	> 45 or $> H/150$

Notes: 1) "δ" is the inter-story deformation or inclination value of wall or column of multistory factory. "h" is the inter-story height of multistory factory.

2) This table is applicable to total building height $H \leqslant 10$ m. When total building height $H > 10$ m, the deformation or inclination value of each level can increase 10% with every 1 m accretion in the total height.

3) Use the lower level of the inter-story deformation and total deformation as the sub-item level of factory deformation.

Table 2.17   Assessment of ground bearing capacity and deformation

No.	Assessment item	a	b	c	d
1	Bearing capacity checking calculation p/s	$\geqslant 1.0$	$<1.0, \geqslant 0.95$	$<0.95, \geqslant 0.90$	<0.90
2	Ground deformation	Deformation has stopped; settlement rate is zero; no overlarge non-uniform settlement	Deformation has stopped generally; settlement rate is less than 2 mm/month in two consecutive months, non-uniform settlement less than standard value of code	Deformation is developing; settlement rate is greater than 2 mm/month in two consecutive months, non-uniform settlement is a little greater than standard value of code; with influence on service of crane, but adjustable	Deformation is developing; settlement rate is greater than 2 mm/month in two consecutive months, non-uniform settlement is greater than standard value of code; with influence on service f crane, and not adjustable

Note: "p" is design value of practical average pressure on bottom surface of foundation or design value of vertical force of single pile. "s" is the design value of ground bearing capacity or vertical bearing capacity of single pile.

Table 2.18   Assessment of function of enclosure structure system

Sub-item name	a	b	c	d
Roof system	Good constitution, well drained	Aged, bubbling, with cracks, slight damage, or clogging, etc., without water leakage	Aged, bubbling, with cracks or corrosion in many places, or with partial damage, perforation, clogging or water leakage	Serious aged, with corrosion or with multiple damage, perforation, cracks in many places, or with severe clogging or water leakage locally
Walls, doors and windows	Intact	Intact walls, doors and windows, with slight damage on finishing, decoration, connecfion and glass, etc.	Partial damage on walls, doors and windows or connection, with influence on serviceability	Severe damage on walls, doors and windows or connection, parts of which have lost function
Underground waterproof	Intact	Generally intact, with humidity in some places, without obvious leakage	With partial damage or water leakage	Damaged in many places, or with serious water leakage
Protective facility	Intact	Slightly damaged, with no influence on protective function	Partially damaged, with influence on protective function	Damage in many places, some parts have lost protective function

Notes: Protection facilities refer to ceilings and various installations for heat insulation, cold insulation, dust-proofing, moisture resistance, corrosion resistance, impact resistance, explosion protection and safety.

Assessment of items or combined items of structures or members is determined according to relevant principles, which are based on such sub-items such as bearing capacity, structural detail, and connection. For example, the assessment of concrete structures or members should be based on bearing capacity, structural detail and connection, crack, and deformation, and be determined as follows:

a) When the differences between the deformation, crack, and the bearing capacity or structural details and connection are within the same grade, the lower grade of bearing capacity or structural details and connectors will be taken as the evaluated grade.

b) When the evaluated grades of deformation and crack are two grades lower than those of bearing capacity or structural details and connectors, decrease one grade of the lower grade of bearing capacity or structural details and connection as the evaluated grade.

c) When the evaluated grades of deformation and crack are three grades lower than those of bearing capacity or structural details and connectors, based on the development speed of deformation and cracks, and their influence on bearing capacity, decrease one or two grades of the lower grade of bearing capacity or structural details and connectors as the evaluated grade.

c. Unit.

Unit is the third level in structural reliability assessment, which refers to the whole or partial building structure, and special structural systems (such as roof system and load-bearing wall). Since assessment of unit, based on the result of assessment of each item, is integral, it is the overall evaluation of the building structure (whole or partial), and expressed as 1, 2, 3, or 4.

(2) Integral assessment

For the structural reliability, the process of integral assessment can be as follows:

a. Division of unit.

Divide the whole, or parts of the building or structure system into one or more units, according to the structural state, structural system, facility layout, working conditions and object of assessment.

b. Assessment of each combined items of unit.

Integral assessment of unit includes load-bearing structure system, structural layout and bracing system, and enclosure structure system. Each category of combined items can be divided into A, B, C, D grades.

c. Integral assessment of unit.

Integral assessment of unit has four grades: load-bearing structure system, structural layout and bracing system, enclosure structure system. It is mainly based on the structure system, and determined as follows:

a) When the evaluated grades of structural layout and bracing system, enclosure structure system are one or less than one grade lower than that of load-bearing structure system, the grade of load-bearing structure system will be taken as the evaluated grade of the unit.

b) When the evaluated grades of structural layout and bracing system, enclosure structure system are two grades lower than that of the load-bearing structure system, decrease one grade of the grade of load-bearing structure system as the evaluated grade of the unit.

c) When the evaluated grades of structural layout and bracing system, enclosure structure system are three grades lower than that of load-bearing structure system, based on principles mentioned above and specific situation, decrease one or two grades of the grade of load-bearing structure system as the evaluated grade of the unit.

d) The importance, durability, and service state of the unit should be taken into consideration in integral assessment, and the assessment result can be adjusted no more than one grade.

d. Integral assessment result.

The form for integral assessment of units of an industrial factory building is shown in Table 2.19.

Table 2.19   Integral assessment of unit of industrial factory building (segment)

Unit	Combined item name	Combined item A, B, C, D	Unit 1, 2, 3, 4	Remark
I	Load-bearing structure system			
	Structural layout and bracing system			
	Enclosure system			
II	Load-bearing structure system			
	Structural layout and bracing system			
	Enclosure system			
...	...			

(3) Principles of grading assessment

Each level can be divided into four grades according to assessment grading standard. Different measures are taken for each grade. The details are shown in Table 2.20.

In this table, when a sub-item such as the bearing capacity of a reinforced concrete structure satisfies current codes, it belongs to rank a; if it fails to meet current codes just a little and belongs to rank b, measures are not needed; if it fails to meet current codes, it belongs to rank c, strengthening is necessary; for total failure to meet current codes (fails to meet the least requirement), it belongs to d, which means it needs to be strengthened, replaced or even discarded immediately.

2. General requirements of structural checking calculation

In the structural reliability assessment, the checking calculation of concrete, steel and masonry structures or members should be in accord with following provisions:

Table 2.20   Grading measures to different grade in detail

Level	Grade	Assessment	Detailed measures
Sub-item	a	Satisfy requirements of current codes, no need of measures	
Item or combined item	A		
Unit	1		
Sub-item	b	Insufficient to meet requirement of current codes, without influence on regular service. Proper measures should be taken for several secondary sub-items or items	Generally only maintenance and some durability treatment or strengthening measures may be needed for some members
Item or combined item	B		
Unit	2		
Sub-item	c	Insufficient to meet requirements of current codes, with influence on regular service. Measures are needed. Proper measures should be taken immediately for several secondary sub-items or items in which the insufficiency is serious	Strengthening needed. Some parts may need replacement
Item or combined item	C		
Unit	3		
Sub-item	d	Failure to meet requirements of current codes is very serious. Accidents may happen at any time. Measures are needed immediately	Measures such as strengthening, replacement or discarding, etc. should be taken immediately
Item or combined item	D		
Unit	4		

a. Checking calculation should be in accord with current codes and standards. Generally, it requires checking calculation in strength, stability, and connecting of structures or members, along with fatigue, crack, deformation, overturn, and slipping if necessary.

For the structures or members which have no definite checking calculation method in current codes or are difficult to be evaluated after checking calculation, considering practical experience and practical structural condition (including assessment test if necessary), they should be evaluated synthetically according to current code *Unified Standard for the Design of Structures*.

b. The calculating sketches of structures or members in checking calculation should be in accord with the actual stress and structural details.

c. Effect on the structure, partial factors and combination coefficient should be determined according to the provisions below. The additional stress due to deformation and temperature should also be considered.

a) The inspected actions on the structure, which conform to the defined value of current national code *Load Codes for Design of Building Structures*, should be determined according to codes; if there is some special situation or no definition in current national codes, it should be determined according to *Unified Standard for the Design of Structures*.

b) The partial factors and combination coefficient of effects should be determined according to current national codes *Load Codes for Design of Building Structures*. If there is sufficient evidence, associating with practical experience, they could be determined through analysis.

d. When the type and performance of material conforms to the original design requirement, material strength used in checking calculation should be the original design value.

When the type and performance of material do not conform to the original design or the material deteriorates after hazard, material strength should be adopted from the tested data.

Standard value of material strength should be determined according to current the national standard *Unified Standard for the Design of Structures*. Probability distribution of material strength ought to be normal distribution or logarithmic normal distribution. Standard value of material strength should be determined by the 0.05 quantile of its probability distribution. When test data is not sufficient, standard values of material performance may be values specified in relevant standards, or determined through analysis, associating with practical experiences.

e. When the surface temperature of concrete structure is higher than 60°C in a long term, or the surface temperature of steel structure is higher than 150°C in a long term, it is necessary to take the temperature influence on material into consideration.

f. Geometric parameters of structures or members in checking calculation should be actual tested values. Each situation should be taken into consideration, such as damage, corrosion, rust-eaten, deviation, weakening of the members and excessive deformation of structures or members.

2.5.2   Building Damage Degree Assessment Method

Building damage degree assessment is a gradation method based on the damage degree of the building. The five grades are: intact buildings, basically intact buildings, generally damaged buildings, seriously damaged buildings, and dangerous buildings. The dangerous buildings are determined by the definition of the dangerous members and dangerous buildings specified in *Standard of Dangerous Building Appraisal*, and the grading of intact buildings, basically intact buildings, generally damaged buildings, seriously damaged buildings is determined according to the *Evaluation Standard of the Building Damage Degree*.

Main jobs in the building damage degree assessment are as follows.

(1) The buildings are divided into the following types according to the structure forms

Reinforced concrete structure-the main bearing structure is constructed with reinforced concrete.

Hybrid structure-the main bearing structure is constructed with reinforced concrete and brick and/or wood.

Brick-wood structure-the main bearing structure is constructed with woods and bricks.

Other structure-the main bearing structure is constructed with bamboo, brick or soil.

(2) Building damage degree standard

There are four grades for the building damage degree: intact, basically intact, generally damaged and seriously damaged. The standard gives the qualifications of intact, basically intact, generally damaged and seriously damaged degree for every kind of structure, decoration and facility (structural components of all kinds of buildings including: foundation, bearing members, non-bearing walls, roofs and floors; decoration including doors and windows, plastering of internal and external walls, ceiling and fine wood decoration; facilities including water supply and lavatory, illuminating system, heating system and other special facilities like fire hydrants and lightening conductors. For example, to reach the standard of "basically intact," the structure should meet the qualifications such as: the foundation of the structure has bearing capacity, with a little uneven settlement that may exceed the limit but has been stable; the bearing members may have a little damage but are basically substantial.

The method of building damage degree evaluation is to evaluate each part seperately according to the damage degree of each part of the building, including structure, decoration and facilities.

Although the *Evaluation Standard of the Building Damage Degree* has specified the qualifications of each damage degree, sometimes in the practice of evaluation, it is difficult to get a conclusion of damage degree without checks or tests for some important and complex buildings or some members with obviously insufficient sections.

There are many details of the damage degree standards in the *Evaluation Standard of the Building Damage Degree*, the details of which are not introduced here.

(3) Method of the damage degree evaluation

According to damage degree standards of the structures (reinforced concrete structure, hybrid structure, brick-wood structure and the other structures), decoration, facilities, and other components, buildings will be graded in four damage degrees: intact, basically intact, generally damaged and seriously damaged buildings.

Table 2.21 shows that this evaluation method is based on the damage degree standard

Table 2.21   Method of the damage degree evaluation of the buildings

Building type	Damage degree	Building damage degree evaluation
Reinforced concrete structures, hybrid structures, Brick-wood structures	Intact buildings	Any building that fulfills one of the following qualifications can be rated as the intact building: 1. Damage degrees of the structure, decoration, and facilities qualify the grade of "intact" 2. One or two damage degrees of the decoration and facilities qualify the grade of "basically intact", while others all qualify the grade of "intact"
	Basically intact buildings	Any building that fulfills one of the following qualifications can be rated as the basically intact building: 1. Damage degrees of the structure, decoration, and facilities qualify the grade of "basically intact" 2. One or two damage degrees of the decoration and facilities qualify the grade of "generally damaged", while others all qualify the grade of "basically intact" 3. One damage degree of the structure (except for foundation, bearing members or the roofs), and one of the decoration or facilities qualify the grade of "generally damaged", while others all at least qualify the grade of "basically intact"
	Generally damaged buildings	Any building that fulfills one of the following qualifications can be rated as the generally damaged building: 1. Damage degrees of the structure, decoration, and facilities qualify the grade of "generally damaged" 2. One or two damage degrees of the decoration and facilities qualify the grade of "seriously damaged", while others all qualify the grade of "generally damaged" 3. One damage degree of the structure (except for foundation, bearing members or the roofs), and one of the decoration or facilities qualify the grade of "seriously damaged", while others at least all qualify the grade of "generally damaged"
	Seriously damaged buildings	Any building that fulfills one of the following qualifications can be rated as the seriously damaged building: 1. Damage degrees of the structure, decoration, and facilities qualify the grade of "seriously damaged" 2. A few of the damage degrees of the decoration and facilities qualify the grade of "generally damaged", while others all qualify the grade of "seriously damaged"
Other structures	Intact buildings	Damage degrees of the structure, decoration, and facilities qualify the grade of "intact"
	Basically intact buildings	Damage degrees of the structure, decoration, and facilities qualify the grade of "basically intact", or only a few of the damage degrees of the building qualify the grade of "intact"
	Generally damaged buildings	Damage degrees of the structure, decoration, and facilities qualify the grade of "generally damaged", or only a few of the damage degrees of the building qualify the grade of "basically intact"
	Seriously damaged buildings	Damage degrees of the structure, decoration, and facilities qualify the grade of "seriously damaged", or only a few of the damage degrees of the building qualify the grade of "generally damaged"

of the building or its members. This method requires entire and particular inspection and research. If necessary, the site test and checking calculation should be done. In conclusion, this method is so direct and convenient that it is widely used.

After the damage degree evaluation of the buildings, the corresponding repair project and retrofitting design should be done.

CHAPTER 3

Retrofitting Design of RC Structures

3.1 Introduction

Reinforced concrete (RC) structures are widely used. Nowadays, many existing RC structures, including some reinforced concrete offices, department stores and plants built in some big cities in China before 1949, have already been repaired. And after the foundation of P. R. China, a great number of civil and industrial buildings and a few office buildings were designed as reinforced concrete structures during the 1950s to 1960s. For those structures of that period, deficiency in construction quality and failure in proper use and maintenance caused problems to their service and even safety. Owing to the expensive cost of structures, it is not economical to replace buildings with new construction because of the insufficient bearing capacity. Retrofit strategies, which require less cost to the buildings, remain the most viable alternative approaches. Additionally, some buildings need to be retrofitted for such requirements such as changing their function or adding new stories. Therefore, retrofit demand will be around for a long period.

Currently, strategies of structural retrofit are various, such as enlarging section area, adding reinforcements, prestress retrofit, changing load path, sticking steel plates and encasing members with steel. Furthermore, chemical grouting should be performed prior to retrofitting with regard to cracked members, and whether to reinforce the members or not will depend on the bearing capacity checking. The biggest difference between structure design and retrofit design is the influence derived from existing members. Therefore, existing conditions, such as actual loading state, surrounding environment, and construction feasibility, should be taken into account substantially as well as safety and economic factors for application of the retrofit strategy adopted in the retrofit design. Hence, it could be unsuitable to certain conditions or structural types for a retrofit strategy whereas effective to other loading states or types. It was found that the strategy of sticking steel was misused in retrofit practice in spite of its applicability. For example, the strategy of sticking steel plates, which is widely used in retrofit practice for its shortcut and convenience in construction, is particularly applicable to flexural members, but unsuitable for axial compression members and small eccentricity members.

Some typical retrofit strategies will be introduced in this chapter in terms of structural types. Also, the design analyses, along with detailing provisions and construction procedures, will be introduced in the sections below for their significance with regard to retrofit construction quality.

3.2 Retrofitting of RC Beams and Slabs

As the most common components, the beams and slabs maintain the largest retrofit workload. The retrofit due to capacity insufficiency involves flexural retrofit of normal section and shear retrofit of inclined section separately, which mainly results from undesirable construction quality, improper design or application, unexpected accident, functional renewal, expiration of durability and so on.

The causes and phenomena of flexural insufficiency in practical retrofitting are introduced in the following sections, and applicable retrofit strategies are proposed based on the cause analysis. The retrofit strategies discussed herein comprise enlarging section area, adding tensile reinforcements, sticking steel plates and prestress retrofit and so on.

These strategies are suitable for varied kinds of members governed by flexural capacity, such as roof beams, floor beams, crane beams, highway bridge beams, frame beams, roof slabs and floor slabs.

3.2.1   Cause and Phenomenon of Capacity Insufficiency of RC Beams and Slabs

Capacity insufficiency of beams and slabs indicates that bearing capacity cannot meet intended demand or requirement for renovated function; therefore member retrofit must be implemented to ensure structural safety. In such cases, the phenomena of capacity insufficiency comprise excessive deflection, over-width of cracks, steel corrosion and concrete crushing in compressed area. In this section, the appearance phenomena and their analyses of capacity insufficient bent members along with damage features of normal sections and inclined sections are summarized below, which will be of benefit to readers to judge whether capacity is insufficient and whether members need to be retrofitted.

1. Causes of capacity insufficiency of RC beams and slabs

The causes result in the capacity insufficiency of RC beams and slabs comprise of some aspects below:

(1) Effect of construction

Insufficient reinforcements and error of reinforcement in construction are regarded as main causes for substandard quality of members. For instance, a garage in Jilin Province, which had RC framing for the first floor and masonry structure for the second floor, was found seriously cracked at the surfaces of beams and slabs. Deflection of a slab reached $L/82$ and the crack width was approximately 1 mm, which gave rise to noticeable vibration even under pedestrian load. Based on the investigation, it was estimated that construction quality was the main cause. Actually, concrete of grade C10 was adopted instead of design grade C20 and it only left 763 mm^2 for reinforcement area of the beam rather than design value of 1251 mm^2. Consequently, the garage was not applicable for use and had to be retrofitted. Another cause was malposition of tensile reinforcements, which could result in cracking of the concrete on tensile side and even rupture of the member, and it occurs more frequently at the ends of cantilever beams or slabs. A case in point would be an accident that happened in Hunan Province. As the design thickness of 100 mm remained 80 mm for balcony slab and the negative moment reinforcements descended by 32 mm, the concrete of top surface at the end of the slab cracked seriously and ultimately the cantilever slab fell down.

In addition, incorrect material application in construction will also debase structural quality and lead to capacity insufficiency of the building. Some typical cases in practice are specified as adoption of moistened or stale cement, employment of plain bars instead of deformed bars, indiscriminate utilization in concrete mix proportion, and application of sand or stone with excessive impurity. As a typical case, the concrete explosion of beams and slabs occurred in several projects after 4 or 5 years of service. Investigation into the cause revealed that harmful impurity of (MgO), which consisted of alkali aggregates or aggregates, formed into [Mg(OH)$_2$] after absorbing water and accordingly gave rise to concrete explosion with rapid expansion.

(2) Effect of design

It is generally believed that the discord between loading status and calculation diagrams and mistakes in calculating loads become the primary design causes for capacity insufficiency. If the secondary beams treated as continuous are calculated as hinge-supported beams to estimate supporting force, the force at middle bearing will absolutely be underestimated by over 20% and the capacity of the main beams will be insufficient consequently. For instance, due to disregarding the self-weight of a brick wall over the secondary beam, shear failure of main beams with small shear span ratio and flexural failure of main beams with large shear span ratio occurred separately in two neighboring masonry buildings. In addition, construction details should also be given enough attention in retrofit design. Local damage of the concrete near anchor zone of prestressed steel is likely to occur if the concrete cast is not compacted enough along with the high density of steel in the anchor zone.

(3) Effect of usage

Overloading in service is another cause for capacity insufficiency of members. For instance, a slag ash roof 100 mm thick was substituted for foam concrete roof of 40 mm thick in an industrial building of Handan City; therefore actual self-weight of the roof increased by 93% after absorbing rainwater, and the roof collapsed as a result of overloading.

Moreover, functional change is another cause for capacity insufficiency of members. Functional changes increase service load significantly and therefore lead to insufficient capacity in slabs and beams, which involve adding or renewal equipment for updating processing technology, augmenting of traffic flow and application of large tonnage truck in bridges, adding stories and renovating function in civil buildings.

(4) Other causes

There are still other factors that account for insufficient capacity.

a. Differential subsidence of foundation results in additional stress of beams.

b. Application of unproved members such as imbrexes may give rise to insufficient capacity. Cracks are often found at internal surfaces of imbrexes and therefore lead to corrosion of tensile reinforcements as well as serious carbonization of concrete covers after 10 years or so. It can result in rupture of the member in the corrosive environment.

c. Effects of member types. Cracks of thin web girders are often found in practice in spite of advances in theory. The inclined cracks shaped like a stone of jujube appear at middle depth and extend rapidly toward both ends at 60%~80% the design value of bearing capacity. Especially under a long-term loading, the cracks become more aggravated. A case was a blacksmith shop built in 1971. Inclined cracks on web surfaces of thin web girders were acquired immediately after completion and consequently developed gradually into compression zones after three months of plastering along with crack width 0.5 mm. In general, excessively thin webs and insufficient web reinforcements as well as lower concrete strength are responsible for inclined cracks of thin webbed girders.

On account of brittle shear failure, retrofit of thin web girder should be done as soon as possible particularly if the inclined cracks are getting wider.

d. Due to durability deficiency, corrosion and rupture of reinforcements generally happen and thereby reduce bearing capacity of the members. A T-shaped beam bridge situated in Fenghua Bridge, Ningbo City was originally built in 1935. The long-term overloading gave rise to subsequent serious crack and even spalling in concrete cover, and it was found that main reinforcements of some bridge beams were corroded to half of the original area and three of them even ruptured, and deflection of a main beam reached 57 mm. Therefore, the bridge had to be retrofitted by prestress strategy in 1981.

Besides the causes discussed previously, insufficient anchor length and lap length, unfirm weld as well as abrupt loading will lead to capacity insufficiency of the members.

2. Damage features of normal section

The cracks of RC flexural members usually appear when forces attain 15%~20% of the ultimate load. Under-reinforced beams appear ductile with increasing loads after cracking and show obvious signs before failure. As for over-reinforced beams whose reinforcement amount is more than the calculated value, the failure is abrupt and cannot be anticipated in advance. In such case, the failure of over-reinforced beams originates from concrete crushing in the compression zone, and the tensile reinforcements, however, are still in elastic range until failure; therefore, the ultimate deflection is negligible in contrast to under-reinforced beams.

Insufficiently reinforced beams, though forbidden in accordance with the code, actually existed due to artificial errors in construction. For instance, misplacement of the reinforcements at the ends of cantilever beams will produce insufficiently reinforced beams and then give rise to abrupt failure.

The load-deflection curves for over-reinforced beams, under-reinforced beams and insufficiently-reinforced beams are respectively plotted in Fig. 3.1. It illustrates that ductility of the member decreases with the increasing of reinforcement ratio ρ. Furthermore, turning points are not visible before failure for over-reinforced beams and insufficiently reinforced beams characterized as brittle failure patterns, distinct from under-reinforced beams. In addition, it can be seen from Fig. 3.1 that sectional resisting moment of L_{3-15} is larger than that of L_{3-14} owing to higher strength of concrete.

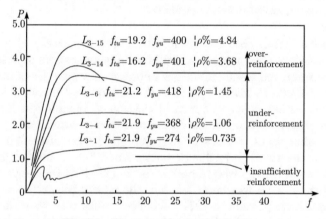

Fig. 3.1   Diagram of load and deflection.

First of all, the original beam should be distinguished among different beam types before retrofitting. Reinforcement ratio ρ of the under-reinforced beam lies between the maximum ratio $\rho_{\max}$ and the minimum ratio $\rho_{\min}$, while that of the insufficiently reinforced beam is smaller than $\rho_{\min}$ and the over-reinforced beam is greater than $\rho_{\max}$.

According to codes, the maximum reinforcement ratio of tensile reinforcement can be expressed as:

$$\rho_{\max} = \xi_b \frac{f_{cm}}{f_y} \tag{3.1}$$

where ξ_b is the coefficient of characteristic compressive height in critical failure, which equals to 0.61 for grade I steel, 0.55 for grade II steel; f_{cm} is the factored concrete flexural compressive strength equals to $1.1f_c$; f_y is the factored tensile strength of the tensile reinforcement.

In accordance with provisions of the code, the minimum ratio of longitudinal reinforcement $\rho_{\min} = 0.15\%$ when concrete grade is under C35.

The insufficiently reinforced beam must be retrofitted using the adding reinforcement method discussed in this section.

With regard to under-reinforced beam, the necessity for retrofit may be judged from crack width, reinforcement stress and deflection of structural members. In general, crack width is a linear function of reinforcement stress, which indicates that the wider the crack, the higher the reinforcement stress. The calculation formula of reinforcement stress in service is given:

$$\sigma_s = \frac{M}{0.87h_0A_s} \tag{3.2}$$

where M is the actual imposed moment; A_s is the total area of longitudinal reinforcement.

Note that over-reinforced beams should be avoided due to adding excessive reinforcements with reference to the adding reinforcement method.

As for over-reinforced beams, the section enlarging method or setting support point should be adopted instead of the strategy of adding reinforcements on tensile surface that is absolutely ineffective.

3. Damage features of inclined section

It is revealed from shear experiments that diagonal cracks involve web-shear diagonal cracks and flexural-shear diagonal cracks. The flexural-shear cracks emerge from extreme tension fibers and extend to diagonal ends by combined action of moment and shear. For thin web beams such as T-shape and L-shape beam, the inclined cracks appear initially near the neutral axis of the web and develop diagonally to both ends of the members.

Generally, the amount of stirrup in a beam has a great effect on failure and shear capacity.

Stirrups could restrict inclined cracks developing and then enhance shear capacity of a beam significantly. With load increasing, one of the diagonal cracks (named critical cracks) expands more rapidly than the others and the intersecting stirrups, once yielding, can hardly restrain the diagonal cracks to shear failure in the compression zone of concrete by combined action of shear and compression, although the load could increase slightly. Accordingly, shear capacity of members depends to a great extent on concrete strength, cross-section dimension, number of stirrups as well as shear span ratio and longitudinal reinforcement ratio.

Excessive stirrups (especially for the thin web beams) prevent critical cracking effectively; however some parallel inclined cracks are observed between stirrups and divide a web into a few inclined compression prisms. Consequently, principal compressive stress of concrete between inclined cracks reaches ultimate strength without stirrups yielding, which leads to diagonal compression failure finally. Shear capacity is determined by section dimension as well as concrete strength.

As for insufficient-stirrup beams, stress of stirrups will enter strain-hardening range immediately after cracking and can hardly resist the tension carried by concrete previously, which results in brittle failure of diagonal tension.

In summary, the number of stirrups decides the features of shear failures that could be corrected based on relevant code.

a. The maximum stirrup ratio could be given as

$$\rho_{sv,\max} = \left(\frac{nA_{sv1}}{bs} \cdot \frac{f_{yv}}{f_c}\right) = 0.153 \tag{3.3}$$

Section dimension should satisfy the requirement of Eq. (3.4) in relation to shear force after transforming.

$$V \leqslant 0.25f_cbh_0 \tag{3.4}$$

where n is the number of stirrup legs; A_{sv1} denotes the section area for single leg stirrup; s represents the spacing of stirrups; h_0 indicates the effective depth of section and b signifies the section width or web thickness; f_{yv} and f_c are specified as tension strength of stirrups and compression strength of concrete respectively.

When the number of stirrups is large enough to satisfy Eq. (3.3) or section dimension of a beam does not meet the requirement of Eq. (3.4), web bars could hardly be utilized fully, which is to say that additional stirrups would not yield until shear failure of the beam. In such case, the retrofit strategy of enlarging section should be selected.

However, the restriction condition should be more rigorous for such members with more severe diagonal cracks such as thin-web girders. Therefore, the Eq. (3.4) is written as

$$V \leqslant 0.2 f_c b h_0 \qquad (3.5)$$

b. The minimal stirrup ratio could be given as

$$\rho_{sv,\min} = \left(\frac{n A_{sv1}}{b h_0} \right)_{\min} = 0.02 \frac{f_c}{f_{yv}} \qquad (3.6)$$

When stirrup ratio is less than the requirement of Eq. (3.6), stirrups yield immediately after cracking and can barely restrain inclined crack development, which leads to diagonal tension failure abruptly. Accordingly, the retrofit technique of adding stirrups should be given first priority on such beams.

Whereas stirrup ratio lies between maximum value and minimal value, the technique of adding web reinforcements may be appropriate for lower value in the range and the technique of enlarging sections with additional stirrups could be favorable provided that the ratio is close to maximum value.

3.2.2 Section Enlarging Method

1. Introduction

The section enlarging method is effective when bearing capacity and member stiffness are far below the code-specified requirement. Furthermore, format of enlarging sections (see Fig. 3.2) may be selected from one-side thickening, two-side thickening or three-side thickening in accordance with loading types, detail features and construction conditions.

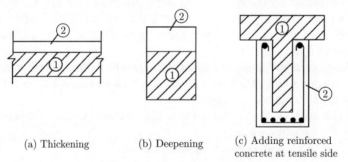

(a) Thickening   (b) Deepening   (c) Adding reinforced
                                                   concrete at tensile side

Fig. 3.2   Section enlarging method:
① original member; ② additional concrete.

When additional concrete is cast above the original concrete surfaces of continuous beams, concrete added at mid-span of beams lies in compression zone, while the concrete is subjected to tensile force at supports. However, thickening underneath is the contrary case.

The concrete added in the tension zone prevents additional reinforcement against erosion, whereas the concrete added in the compression zone enhances effective depth of the section, and increases stiffness and strength of the member. The retrofit strategy of adding concrete layer is definitely feasible.

Generally, the concrete is often added at the tension zone in practice. It is regarded as an effective method to improve flexural capacity especially for a T-shaped beam with low reinforcement ratio and short compression depth by means of adding tensile reinforcements and casting concrete subsequently (see Fig. 3.2). Similarly, reinforcements and concrete should be added above original layers in the case of balconies, canopies and cornice slabs.

The actual compressive strength of concrete for existing members should not be less than 13 MPa.

Stress level of existing members (β_k) should be checked in the retrofit design. It is suggested to perform unloading when the value of critical stress cited in Table 3.1 cannot be satisfied.

Table 3.1   Critical Value of Stress Level for Existing Members [β_k]

Stress Type	Critical Value of Stress Level
Axial Compression and Shear of Diagonal Section	0.70
Compression with Small Eccentricity, Torsion and Partial Compression	0.80
Compression with Large Eccentricity, Flexure, Tension and Partial Tension	0.90

Notes: Where deflection and crack width do not exceed critical value of the code, the value of β_k may increase by 10%, but less than 0.95.

Co-working behavior of additional concrete and original concrete is of significance for composite members with regard to the strategy of enlarging the section and can be ensured based on necessary construction details and techniques. Therefore, detailed measurements and rational procedures are essential to ensure the mechanical behavior of the retrofitted members.

a. First, roughen the concrete surface of the existing member for desirable bonding behavior. In such cases, surface roughness of slabs should not be less than 4 mm while that of beams should not be less than 6 mm along with a notched groove to form concrete shear studs for certain spacing.

b. Next, coat the roughened surfaces with acrylic cement paste (or 107 polymer cement paste) formed by mixing cement with polymer after flushing them thoroughly; finally cast fresh concrete onto the surface. Generally, strength of the acrylic cement paste is $2 \sim 3$ times as high as that of general-purpose slurry, and the mixing ratio as well as product property are clarified in usage instructions and relevant reference material. 107 polymer cement paste is made by adding 107 glue into cement and mixing well.

c. In addition, it is required to configure stirrups and negative moment reinforcements after surface disposal and connection details should also be given much attention. In such cases, additional reinforcements should be welded to original reinforcements with stub bars especially at the longitudinal ends.

U-shaped stirrups shown in Fig. 3.3(a) may be welded to original stirrups and weld length should not be less than 5 d (d represents diameter of U-shaped stirrups). Furthermore, Fig. 3.3(b) shows detailed measurements of the connections between U-shaped stirrups and anchor bolts. In accordance with the provisions of the code, diameter of anchor bolts should

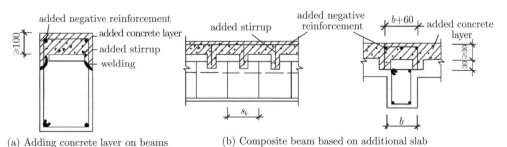

(a) Adding concrete layer on beams   (b) Composite beam based on additional slab

Fig. 3.3   Detailing of adding concrete on beams.

not be less than 10 mm and the distance from the centroid of the bolt to the edge of member should not be less than 3 d and 40 mm. Also, anchor depth of the bolts should not be less than 10 d and the anchor bolts should be anchored with epoxy mortar or epoxy grout to the hole drilled in existing beams. The diameter of the anchor hole should be 4 mm larger than that of anchor bolts. In addition, U-shaped stirrups may as well be anchored into anchor holes directly rather than by means of anchor bolts.

2. Calculation method

(1) Mechanical performance

Mechanical performance of the retrofitted members for every loading stage is illustrated in Fig. 3.4. It is evident that moment M_1 has already existed in existing members before casting additional layers, which gives rise to sectional stress shown in Fig 3.4(b) and the stage is called the first phase of loading. As long as the strength of additional concrete reached the design strength, both parts of the retrofitted members initiate to resisting following moment M_2 together. The incremental moment M_2 is resisted by the whole depth of the retrofitted members h_1 and the sectional stress indicated in Fig. 3.4(c), which is entitled second phase of loading. Accordingly, total stress of the section by the action of M_1 and M_2 is displayed in Fig. 3.4 (d).

It is indicated from Fig. 3.4(d) that sectional stress obtained from secondary loading differs from that of first phase of loading. The difference in two aspects is explained below.

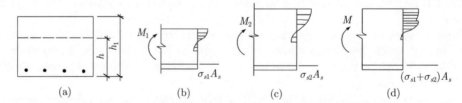

Fig. 3.4   Mechanical behavior of retrofitted section.

a. Strain lag of concrete.

Compared with ordinary concrete beam, the additional concrete in retrofitted beam participates in flexural resistance after the action of moment M_1 undertaken by original concrete. Consequently, compression strain of the additional concrete is always smaller than that of the original concrete, which is called strain lag of concrete.

Deflection levels and crack widths of retrofitted members are definitely larger than those of ordinary concrete beams when the compress zone concrete is crushed.

b. Stress advance of reinforcement.

In the first phase of loading, reinforcement stress σ_{s1} and deflection f_1 for original members are larger than those of corresponding ordinary concrete beams at the same action of moment M_1 due to the small section depth. When moment M_2 is imposed, location of neutral axis of the retrofitted section moves up and part of the compression zone in first phase of loading turns to tension state in second phase of loading. In other words, the compressive pressure could be regarded as prestress for second phase of loading, which is called "loading prestress". Loading prestress can decrease the reinforcement stress and deflection under the action of moment M_2. However, because of the action of M_1, the total value on reinforcement stress, deflection and crack width is much larger than that of ordinary concrete members for entire phases, and the tensile reinforcement could reach yield limit under much lower moment action, which is called stress advance of reinforcement.

(2) Calculation and control of reinforcement stress in service

Stress may cause deflection and crack width to exceed code-specified requirement; tensile stress of steel stays higher and the steel even reaches yield limit. Therefore, it is essential for retrofit design to calculate the stress of reinforcements in retrofitted members and make sure it is lower than code-specified requirement during service. Service stress of the tension reinforcement of retrofitted members is given by

$$\sigma_s = \sigma_{s1} + \sigma_{s2} \leqslant 0.9 f_y \tag{3.7a}$$

where σ_{s1} is reinforcement stress due to moment M_1 before additional concrete active, and can be determined below:

$$\sigma_{s1} = \frac{M_1}{A_s \eta_1 h_0} \tag{3.7b}$$

σ_{s2} is reinforcement stress due to moment M_2 after additional concrete active and can be determined below:

$$\sigma_{s2} = \frac{M_2(1 - \beta)}{A_s \eta_2 h_{01}} \tag{3.7c}$$

where A_s is sectional area of the tensile reinforcement; η_1 and η_2 represent internal lever arm coefficient at the cracked section and both may be taken as 0.87 approximately; β is composite characteristic parameter and may be taken in accordance with Eq. (3.7d), which indicates the effect of load prestressing and the ratio of h_1/h.

$$\beta = 0.5 \left(1 - \frac{h}{h_1}\right) \tag{3.7d}$$

f_y denotes design strength of tension reinforcement; h and h_0 are sectional depth and effective depth of the section for original members; also, h_1 and h_{01} are sectional depth and effective depth of the section for retrofitted members respectively.

(3) Calculation of bearing capacity

In the retrofit by section enlarging method, the estimation of bearing capacity should conform to the provisions of the *Code for Design of Concrete Structure (GB50010—2002)* in China and much attention should be given to the interaction between additional concrete and original concrete.

When the strategy of section enlarging method is adopted for retrofit of bent members, the retrofit design should conform to the provisions of the *Code for Design of Concrete Structure (GB50010—2002)* and *Technical Specification for Strengthening Concrete Structures (CECS 25:90)* in China.

a. The depth of compression zone of relative boundary value ξ_b could be determined as follows with regard to enlarged section.

a) As to enlarging section on one side of additional steel at tensile surface

$$\xi_b = \frac{0.8}{1 + \left[\dfrac{f_y}{E_s} + \dfrac{\sigma_{s0}}{E_{s0}}\left(1 + \dfrac{2\delta}{h_{01}}\right)\right] \cdot \dfrac{1}{0.0033}} \tag{3.8}$$

b) As to enlarging section at compressive surface

$$\xi_b = \frac{0.8}{1 + \dfrac{f_y}{E_{s0}\left[0.0033 + \dfrac{\sigma_{s0}}{E_{s0}}\left(1 + \dfrac{2\delta}{h_{01}}\right)\right]}} \tag{3.9}$$

c) As to enlarging section at both surfaces and neglecting discrepancy on thickness for both surfaces

$$\xi_b = \frac{0.8\left[0.0033 + \dfrac{\sigma_{s0}}{E_{s0}}\left(1 + \dfrac{2\delta}{h_{01}}\right)\right]}{0.0033 + \dfrac{f_y}{E_s} + 2\dfrac{\sigma_{s0}}{E_{s0}}\left(1 + \dfrac{2\delta}{h_{01}}\right)} \tag{3.10}$$

where f_y is design strength of additional reinforcement; E_s is elastic modulus of additional reinforcement; σ_{s0} is stress of reinforcements which lies at the side with lower stress level for original part while retrofitting; E_{s0} is the elastic modulus of original reinforcements; δ is the thickness of additional layer and could be taken as the average thickness of double surfaces; h_{01} is the effective depth of original section.

b. When additional concrete is cast onto compressive surfaces of flexural members, loading effect should be determined in accordance with the provisions of *Code for Design of Concrete Structure*.

c. Regarding enlarging sections with additional reinforcements at tensile surface, design strength of the reinforcements should be multiplied by strength utilization coefficient 0.9 for flexural calculation.

d. When calculating the shear capacity of inclined sections, the strength utilization coefficient for additional concrete and additional reinforcement should be taken as 0.75 and 0.85 separately and the contribution of unanchored U-shaped stirrups for shear capacity should be ignored.

3. Detail provisions

a. The minimal thickness of additional concrete should not be less than 40 mm for slabs, 60 mm for beams, and 50 mm for shotcrete construction.

b. Pebble and gravel may be used as coarse aggregate for their hardness and durability, and the maximum particle size may not be more than 20 mm.

c. Diameter of the additional reinforcements may be 6~8 mm for slabs and 12~25 mm for beams, and deform steel is preferred for longitudinal reinforcement of the beams. Also, diameter of the enclosed stirrups may not be less than 8 mm and diameter of the U-shaped stirrups may be the same as the original stirrups.

d. The net spacing between original reinforcements and additional reinforcements should not be less than 20 mm and welded stub bars should be employed for their connection; furthermore, enclosed stirrups or U-shaped stirrups should be adopted in accordance with the *Code for Design of Concrete Structure*.

a) When stub bars are employed to connect additional reinforcements and original reinforcements (see Fig. 3.5(a)), it is suggested that diameter and length for the bars shall not be less than 20 mm and 5 d (d is the lesser value of the diameter for original reinforcements and additional reinforcements) and the spacing between stubs shall not be more than 500 mm.

b) U-shaped stirrups should be adopted for double-surface retrofit as well as single-surface retrofit (see Fig. 3.5(b)).

U-shaped stirrups may be welded to original stirrups and minimal weld length should reach 10 d and 5 d separately for one side welding and double side welding (d is the diameter of U-shaped stirrups).

Also, stirrups could be anchored to anchor holes directly as well as be welded to steel bars or anchor bolts which are anchored in existing members (see Fig. 3.5(c)). The diameter of steel bars or anchor bolts should not be less than 10 mm and distance from centroids of the bolts to the edges of members should not be less than 3 d and 40 mm. Furthermore, anchor depth of the bolts should not be less than 10 d and epoxy mortar or epoxy grout should be used to anchor the bolts to the holes drilled in existing beams. In addition, the diameter of anchor holes should be 4 mm larger than that of anchor bolts, and anchor behavior, especially at the longitudinal ends, should be validated for reliability.

e. Both ends of the added reinforcements should be anchored reliably.

4. Construction requirements

a. The procedures and provisions below should be followed in retrofit of RC structures.

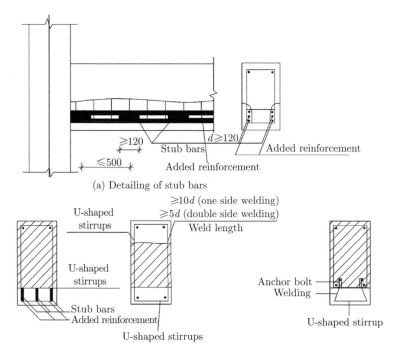

(a) Detailing of stub bars

(b) Welding U-shaped stirrups to existing stirrups   (c) Welding U-shaped stirrups to anchor bolts

Fig. 3.5   Details of connection between additional reinforcements and existing member.

a) Chisel off plaster layer until the structural layer is reached and toughen the surfaces or notch grooves for certain spacing. As to the grooves, depth may not be less than 6 mm and spacing may not be larger than 200 mm as well as the spacing of additional stirrups. Furthermore, concrete edges of existing members should be filleted and dregs should be removed as well.

b) Flush the roughened surface of existing members thoroughly and cover it with fresh cement paste or interfacial agent before casting additional concrete.

b. The next procedure is to remove the rust on reinforcements. Also, it is suggested that such a measurement as unloading or set bracing shall be executed after derusting and then reinforcement shall be welded rebar after rebar, region after region, section after section, layer after layer gradually to reduce the effects of the created heat as much as possible.

c. In addition, the procedures including supporting formwork, banding reinforcement, casting concrete, and curing concrete should meet the Code for Acceptance of Constructional Quality of Concrete Structures.

5. Calculation example

Example 3.1   A cast-in-situ multi-stories RC frame, which is used as a parking garage for the first floor and an office above the second floor, carries live load 2 kN/m^2. As a result of renovation, the live load of the second floor increases to be 4 kN/m^2. As a result, it is found that floor slabs are insufficient with regard to bearing capacity and should be prepared to retrofit by the strategy of enlarging section.

Solution:

a. Known data: The concrete has grade C20 and the steel has grade 235. Structural thickness of the existing slabs is 70 mm and cover thickness of the cement plaster is 20 mm; Specific dimension and steel configuration appear in Fig. 3.6.

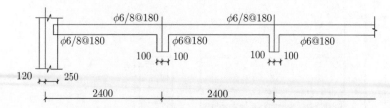

Fig. 3.6   Reinforcement amount of existing slab.

b. Retrofit procedures: Firstly roughen the surface of existing slabs to roughness more than 4 mm for favorable interaction behavior, and then notch grooves 30 mm wide and 10 mm deep as concrete shear studs for every 500 mm. Finally flush the roughened surface thoroughly and cast additional concrete. According to detailing provisions, the thickness of the additional layer is taken to be 40 mm, so the total thickness of the retrofitted slab is to be 130 mm thick (including existing slab 70 mm thick and cover layer 20 mm thickness).

c. Load calculation:

Dead load of the existing slabs	2250 N/m^2
Dead load of the additional slabs	$25000 \times 0.04 = 1000 \text{ N/m}^2$
	$q_1 = 3250 \text{ N/m}^2$
Live load	$q_2 = 4000 \text{ N/m}^2$
Total load	$q = q_1 + q_2 = 7250 \text{ N/m}^2$

Calculation is illustrated in Fig. 3.7.

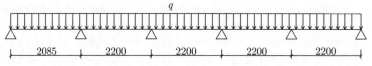

Fig. 3.7   Calculation diagram of the slab.

d. Calculation of internal force.

Total moment for every section is listed in Table 3.2.

Table 3.2   Calculation of Slab Moment

Sections	Middle Section of Side Span	Supported Section of Side Span	Middle Section of Middle Span	Supported Section of Middle Span
Moment M_1 from q_1	1284	−1043	983	−983
Moment M_2 from q_2	1581	−1284	1211	−1211
Total Moment $M_z = M_1 + M_2$	2865	−2327	2194	−2194

e. Bearing capacity of sections.

Bearing capacity of the sections is listed in Table 3.3. Note that $h_1 = 130$ mm and $h_{01} = 115$ mm are for middle sections and $h_1 = 130$ mm corresponds to supported sections. Also, effective depth of the negative reinforcements is defined as 55 mm and 115 mm respectively for existing slabs and retrofitted slabs.

Table 3.3   Validation of the Bearing Capacity

Sections	Middle Section of Side Span	Supported Section of Side Span	Middle Section of Middle Span	Supported Section of Middle Span
$\xi = \rho \dfrac{f_y}{f_{cm}}$	0.036	0.076	0.026	0.055
Solution for α_s	0.039	0.079	0.026	0.057
$M_u = \alpha_s f_{cm} bh_0^2$	5674	−2629	3782	−1897
Comparison with M_z	$M_u \gg M_z$	$M_u > M_z$	$M_u \gg M_z$	$M_u < M_z$

It is evident from Table 3.3 that the flexural capacity is inadequate and additional re-inforcements are required with regard to supported section of middle span, yet minimal ratio of tensile reinforcement ($\phi6@200$) must be sufficient for the negligible insufficiency ($M_z - M_u = 2194 - 1897 = 297$ N $\cdot$ m).

f. Calculation and control of reinforcement stress.

Reinforcement stress σ_{s1} and σ_{s2} could be determined in accordance with Eq. (3.7), and for the middle section of side span, the reinforcement stress is given as

$$\sigma_{s1} = \frac{M_1}{0.87 A_s h_0} = \frac{1284000}{0.87 \times 218 \times 55} = 127 \text{ MPa}$$

$$\beta = 0.5 \left(1 - \frac{h}{h_1}\right) = 0.5 \times \left(1 - \frac{70}{130}\right) = 0.231$$

$$\sigma_{s2} = \frac{M_2(1 - \beta)}{0.87 h_{01} A_s} = \frac{158 \times 10^4 \times (1 - 0.231)}{0.87 \times 218 \times 115} = 55.7 \text{ MPa}$$

$$\sigma'_s = \sigma_{s1} + \sigma_{s2} = 127 + 55.7 = 182.7 \text{ MPa} < 0.9 f_y = 189 \text{ MPa}$$

In addition, for supported section, because additional concrete lies on the tensile surface, it does not have leading stress effect in reinforcements as mentioned above. In contrast, the ad-ditional reinforcements have the effects of stress lag compared with existing reinforcements. Therefore, it is generally accepted that stress validation of additional reinforcements at sup-ported sections of continuous beams is not necessary. The retrofitted slabs are illustrated in Fig. 3.8.

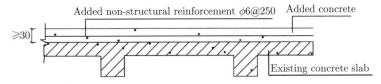

Fig. 3.8   Construction drawing of retrofit strategy of adding concrete layer.

3.2.3   Retrofitting by Adding Tensile Reinforcement

Retrofitting with tensile reinforcement is to add reinforcements on the tensile side of the beam to enhance its bearing capacity. The method is applicable where stiffness of the beam section and the shearing capacity are enough but the tensile strength of the bend region is insufficient and the adding reinforcement is not excessive. In this section, the procedure, characteristics and construction of retrofitting with adding tensile reinforcement, as well as the calculation method of the bearing capacity of the retrofitted members, will be introduced.

1. Introduction of retrofitting with adding tensile reinforcement

Fig. 3.9 shows the retrofitting with adding tensile reinforcement. The connection between additional reinforcements and existing beams involves three forms: full welding, semi-welding and bonding connections.

(1) Full welding

Regarding full welding, additional reinforcements are directly welded to original reinforce-ments, and additional concrete layers are not necessary. The additional reinforcements are exposed to natural environment and participate in flexural resistance with original reinforce-ments by the action of weld (see Fig. 3.10).

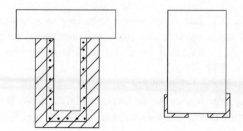

Fig. 3.9   Retrofitted beam of adding reinforcement.

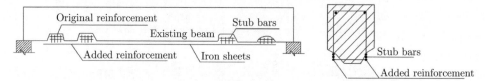

Fig. 3.10   A beam retrofitted by full welding.

In general, it is desirable to locate welding to the inflection points of existing beams where tensile stress of the original reinforcements is negligible. Therefore, concentrated load transmitted through weld from additional reinforcements may result in limited influence on original reinforcements that could be regarded as anchors for additional reinforcements.

(2) Semi-welding

Semi-welding casts fine-grained concrete layers after welding. Additional reinforcements definitely benefit from bonding with additional concrete as well as the anchor to original reinforcements. Therefore, mechanical behavior of the additional reinforcements is almost the same as that of original reinforcements and the reliability of retrofitted members has been improved further.

(3) Bonding concrete

The bonding concrete technique denotes that additional reinforcements contribute to flexural capacity of retrofitted members based only on the bonding strength of concrete.

Construction procedures are introduced herein.

a.   Roughen the surfaces of existing members to ensure surface roughness more than 6 mm.

b.   Notch a groove for every 500 mm as a concrete shear stub and then weld U-shaped stirrups to original reinforcements or to anchor bolts for favorable mechanical behavior.

c.   Thread longitudinal reinforcements into U-shaped stirrups and then bind them together. Finally smear epoxy adhesive onto the interface before casting or ejecting additional concrete.

2. Mechanical performance

It has been proven that stress lag will result in later yield for additional reinforcements and more noticeable deflection and cracks when additional reinforcements are yielding. Generally speaking, the primary cause for stress lag is undischarged load while retrofitting, including self-weight of the members, which gives rise to the initial stress of original reinforcements. In addition, it exerts considerable effect on stress lag for local deformation in welding zone, initial flatness of additional reinforcements, shear deformation and slip of interfaces, initial gaps between steel jackets and beams, local deformation at anchor points, etc.

Local flexural deformation (see Fig. 3.11(b)) of original reinforcements at both sides of welding points is another mechanical feature for the strategy of adding reinforcements. Due

to the stress difference of sections as well as the eccentricity e_0 between additional reinforcements and original reinforcements (see Fig. 3.11(a)), additional moment and induced
flexural deformation are obtained.

The flexural deformation not only aggravates stress lag of additional reinforcements but
causes asymmetrical stress of original reinforcements at both sections of welding points,
which should be given close attention in retrofit design.

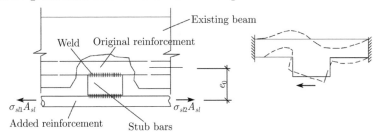

Fig. 3.11   Local flexural deformation of original reinforcements at both sides of welding points.

3. Retrofit design of beams

(1) Calculation of bearing capacity
Based on previous analyses, it is clear that stress of additional reinforcement lags behind
that of original reinforcements; therefore design strength of additional reinforcement should
be multiplied by reduction factor 0.9.

$$\left.\begin{array}{l} f_{cm}bx = f_y A_s + 0.9 f_{y1} A_{s1} \\[2mm] M_u = f_{cm}bx \left(h_{01} - \dfrac{x}{2} \right) \end{array}\right\} \tag{3.11}$$

Eq. (3.11) may be rewritten as

$$\alpha_s = \frac{M}{f_{cm}bh_{01}^2} \tag{3.12}$$

where α_s is coefficient of sectional resistance moment. Provided that the internal lever arm
γ_s is obtained from α_s by design chart, the cross-section area needed could be given as

$$A_{s1} = \frac{M - f_y A_s \gamma_s b_0}{0.9 f_{y1} \gamma_s h_{01}} \tag{3.13}$$

where f_{cm} is design value of flexural compressive strength of concrete and may be taken
as $1.1 f_c$; x is concrete compression height; f_y and f_{y1} are design value of tensile strength
of original reinforcements and additional reinforcements respectively; A_s and A_{s1} are sectional area of original reinforcements and additional reinforcements; h_{01} is effective depth
of retrofitted section, which is the distance from extreme compression fiber to the point
of resultant forces of the original reinforcements and additional reinforcements, and could
roughly be substituted for effective depth of existing beams with regard to small numbers
of additional reinforcements. M represents design value of the moment and M_u denotes the
design values of flexural capacity of retrofitted beams.

Application range of Eqs. (3.11) to (3.13) is the same as that for under-reinforced beams
specified in the code.

(2) Calculation and control of steel stress in service
Stress lag of additional reinforcements is likely to induce the result that original reinforcements enter strain-hardening range prior to additional reinforcements; deflection and
crack width are much larger than those of primary loading members at ultimate load. It is
indicated that stress lag will give rise to higher service stress for original reinforcements and

even lead the steel to liquid limit. Therefore, service stress of original reinforcements should be checked as

$$\sigma_s = \sigma_{s1} + \sigma_{s2} \leqslant 0.8 f_y \tag{3.14}$$

where σ_{s1} is the initial stress of original reinforcements due to moment M_{1k} obtained from undischarged load before casting additional concrete and could be determined by Eq. (3.15a); σ_{s2} is the incremental stress of original reinforcements due to incremental moment M_{2k} after casting additional concrete and could be determined by Eq. (3.15b).

$$\sigma_{s1} = \frac{M_{1k}}{0.87 A_s h_0} \tag{3.15a}$$

$$\sigma_{s2} = \frac{M_{2k}}{0.87 (A_s + A_{s1}) h_{01}} \tag{3.15b}$$

If Eq. (3.14) cannot be satisfied, it is essential to add temporary prestressed struts to impose an opposite load and thereby reduce stress σ_{s1} of original reinforcements before retrofitting.

In addition, sectional capacity of original reinforcements at welding point should be checked for the technique of full welding. In such case, sectional stress of original reinforcement at outer sections of welding points may be satisfied as

$$\sigma_s \leqslant 0.7 f_y \tag{3.16}$$

where σ_{s2} is sectional stress of the original reinforcement at the outer section of welding point and could be determined by Eq. (3.15a) in which the moment means the action of entire load for calculational section.

4. Detailing provisions

Regarding retrofit strategy of adding tensile reinforcement for flexural members, the subsequent provisions should be satisfied.

a. The diameter of additional reinforcements should be chosen in the range of 12~25 mm and ribbed reinforcements are preferred when the technique of bonding concrete to original surfaces was adopted to transmit stress of the additional reinforcements.

b. Net spacing between original reinforcements and additional reinforcements should not be less than 20 mm; diameter of additional reinforcements, with regard to connection with welded stub bars, should not be less than 20 mm and length of the stub bars should not be less than 5 d (d is the less value of diameter for original reinforcements and additional reinforcements) but not more than 120 mm. Furthermore, the spacing of stub bars may not be more than 500 mm for the segments of large gradient moment, yet the requirement may be looser for a segment with small gradient moment. However, number of welds along each of the reinforcements should not be less than four.

c. Diameter of additional reinforcements should be 4 mm less than that of original reinforcements with regard to the technique of full welding.

d. The interface to be bonded should be roughened when the stress of additional reinforcements is transmitted by bonding of additional concrete; also the roughness of the interface should not be less than 6 mm and a concrete groove 70 mm × 30 mm acts as a shear stud may be notched for every 500 mm. Diameter of U-shaped stirrups may not be less than 8 mm, and it could be referred to relevant entries in the retrofit strategy of enlarging sections for connection techniques and detailing provisions between U-shaped stirrups and existing beams. Furthermore, gravel may be adopted for coarse aggregate of additional concrete for its hardness and particle size may not be more than 20 mm. In addition, strength grade of additional concrete composed of 525# Portland cement should be one rank higher than that of original concrete and shotcrete construction may be favorable.

5. An example of engineering interest

Example 3.2   A simply supported RC beam, with calculational span $l_0 = 3.2$ m and section size 150 mm $\times$ 300 mm, has longitudinal reinforcements 2Φ16 and stirrups $\phi6@150$. The beam carries dead load 15.6 kN/m and live load 4.4 kN/m. However, live load increases to 6 kN/m^2 for improved utilization function. Calculate the retrofit.

Solution:

a. Check flexural capacity of the section.

$$M = \frac{1}{8} \times 1.2 \times 15.6 \times 3^2 + \frac{1}{8} \times 1.4 \times (4.4 + 6) \times 3^2 = 37.44 \text{ kN} \cdot \text{m}$$

Compression depth of the original section (corresponds to 2Φ16)

$$x = \frac{f_y A_s}{f_{cm} b} = \frac{310 \times 402}{11 \times 150} = 76 \text{ mm}$$

$$M_u = f_{cm} b x \left(h_0 - \frac{x}{2} \right) = 11 \times 150 \times 76 \times \left(265 - \frac{76}{2} \right) = 2.85 \times 10^7 \text{ N} \cdot \text{mm}$$

$$= 28.5 \text{ kN} \cdot \text{m} < 37.44 \text{ kN} \cdot \text{m}$$

(Insufficient flexural capacity, need to retrofit)

b. Retrofit procedures.

As reinforcement ratio of the existing beam is fairly small, retrofit strategy of adding reinforcement will be beneficial and then the technique of full welding is adopted for connecting. The construction procedure is given below. Concrete cover near welding point is chiseled and then temporary struts are supported in the middle of the span before gradual welding.

c. Calculation of flexural capacity of normal section.

It could be attained from Eq. (3.12).

$$\alpha_s = \frac{M}{f_{cm} b h_{01}^2} = \frac{37.44 \times 10^6}{11 \times 150 \times 265^2} = 0.373$$

It gives $\gamma_s = 0.797$ by referring to design table.

$$A_{s1} = \frac{M - f_y A_s \gamma_s h_0}{0.9 f_{y1} \gamma_s h_{01}} = \frac{37.44 \times 10^6 - 310 \times 402 \times 0.797 \times 265}{0.9 \times 310 \times 0.797 \times 265} = 188 \text{ mm}^2$$

Select $2\phi12(A_{s1} = 226 \text{ mm}^2)$.

d. Calculation of shear capacity of diagonal section.

$$V = \frac{1}{2} \times 1.2 \times 15.6 \times 3 + \frac{1}{2} \times 1.4 \times (4.4 + 6) \times 3$$

$$= 49.92 \text{ kN} < 0.25 \, f_c b h_0 = 0.25 \times 10 \times 150 \times 265 = 99375 \text{ kN}$$

(Sectional dimension satisfies the requirement)

$$V_{cs} = 0.07 f_c b h_0 + 1.5 f_y \frac{A_s}{S} h_0 = 0.07 \times 10 \times 150 \times 265 + 1.5 \times 210 \times \frac{28.3 \times 2}{150} \times 265$$

$$= 278525 + 31498 = 59323 \text{ N} = 59.323 \text{ kN} > 49.92 \text{ kN}$$

(Shear capacity of diagonal section satisfies requirement)

e. Check on service stress of original reinforcements.

When dead load is imposed on existing beam during construction, the characteristic value of the moment is

$$M_{1k} = \frac{1}{8} \times 15.6 \times 3^2 = 17.5 \text{ kN} \cdot \text{m}$$

$$\sigma_{s1} = \frac{M_{1k}}{0.87 A_s h_0} = \frac{17.5 \times 10^6}{0.87 A_s h_0} = 189 \text{ N/mm}^2$$

After retrofitting, the incremental moment is

$$M_{2k} = \frac{1}{8} \times (4.4 + 6) \times 3^2 = 11.7 \text{ kN} \cdot \text{m}$$

$$\sigma_{s2} = \frac{M_{2k}}{0.87 (A_s + A_{s1}) h_{01}} = \frac{11.7 \times 10^6}{0.87(402 + 226) \times 265} = 80.8 \text{ N/mm}^2$$

In accordance with Eq. (3.14), the total stress is

$$\sigma_s = \sigma_{s1} + \sigma_{s2} = 189 + 80.8$$
$$= 269.8 \text{ N/mm}^2 > 0.8 \, f_y = 248 \text{ N/mm}^2$$

It is concluded from the calculation that the stress of original reinforcement in the middle section of the span is so high that the requirement could not be met. Therefore, subsequent measurements are compulsory. Two pieces of temporary bracing rods are set at one-third points of the span before being jacked against the beam, which is achieved by a steel triangular prism wedged between bracing rods and the beam. Assuming that the prestressing force is 15 kN, M_{1k} changes to

$$M_{1k} = 17.50 - 1.0 \times 15 = 2.5 \text{ kN} \cdot \text{m}$$

$$\sigma_{s1} = \frac{2.5 \times 10^6}{0.87 \times 402 \times 265} = 27 \text{ N/mm}^2$$

Two forces of 15 kN are imposed reversely at the one-third points of the span after welding and removing temporary rods. Therefore, M_{2k} could be written as

$$M_{2k} = \frac{1}{8} \times (4.4 + 6) \times 3^2 + 1.0 \times 15 = 26.7 \text{ kN} \cdot \text{m}$$

$$\sigma_{s2} = \frac{26.7 \times 10^6}{0.87 \times (402 + 226) \times 265} = 183 \text{ N/mm}^2$$

$$\sigma_s = 27 + 183 = 210 \text{ N/mm}^2 < 248 \text{ N/mm}^2$$

It satisfies the requirement.

Finally, Eq. (3.16) is employed to validate the stress of original reinforcements at the outer section of welding point; the characteristic value of moment is given as

$$M_k = \frac{1}{2}(15.6 + 4.4 + 6) \times 3 \times (0.4 - 0.12) = 10.92 \text{ kN} \cdot \text{m}$$

$$\sigma_s = \frac{M}{0.87 \, A_s h_0} = \frac{10.92 \times 10^6}{0.87 \times 402 \times 265} = 118 \text{ N/mm}^2 < 0.7 \, f_y = 217 \text{ N/mm}^2$$

It also satisfies the requirement.

3.2.4   Prestress Retrofitting Method

Prestress retrofitting utilizes prestressed reinforcement to retrofit beams or slabs of buildings, which is not only easy to construct, but also able to improve the bending resistance, shear strength and performance of beams and slabs without increasing their section heights and reducing the structure's headroom. The merit of prestress retrofitting is mainly that the negative moment caused by prestress counteracts a part of loading moment, resulting in diminishing the moment of beam and slab; the crack width can also be lessened or even closed.

It is better to retrofit beams by using curved prestressed reinforcement. Therefore, the pre-stress retrofitting method is widely used in beam retrofitting. For example, an 'I' shape beam of a bridge with 20 m span had a deflection of 5.4 cm at middle of span and widest crevice of 0.5 mm before being retrofitted; after it was retrofitted by $4\phi25$ lower-supported prestressed reinforcements, the largest span was not only offsetting the load deformation, but also upwarp for 0.47 cm. The load of two arrays of car-10 was eccentrically applied before being retrofitted, while the load of two arrays of car-13 can be eccentrically applied after being retrofitted. For another example, a lot of cracks appeared in a thin webbed roof beam of a factory building after one year service; one of the thin web beams had more than 63 items of crack; some cracks were developed into the whole web height, and the largest crack width even reached up to 0.6 mm. The reasons for this phenomenon were too thin (100 mm) web, too little steel in the web and too low strength of concrete. After being retrofitted by lower-supported prestressed reinforcements, oblique cracks and vertical cracks were closed and the beams worked well.

The prestress retrofitting method, effect of prestress, calculation of bearing capacity of retrofitted components and numeration of stretch elongation are illustrated in this section. The discussion below is for beam components; however, the principles and methods are also applicable to plates.

There are two working types of prestressed reinforcements. The first one is to deliver force through anchoring point and supporting point externally to an original beam, which is simple, effective and widely used in engineering; the second one is to deliver force through bonding effect between the new and the old concrete by pouring concrete after stretching prestressed reinforcements. The construction and calculation techniques of external prestress retrofitting method are listed below.

1. Prestress retrofitting techniques

The basic techniques of retrofitting beams and slabs by prestressed reinforcements are:

a. Add prestressed reinforcements in the external tensile region that is in need of being retrofitted.

b. Stretch prestressed reinforcements and anchor them at the ends of beams (slabs).

The stretching method and anchoring techniques of prestressed reinforcements are de-picted below.

(1) Stretching prestressed reinforcements

The prestressed reinforcements which are used to retrofit beams are usually put in the externals of beams, so the stretching process is done. There are various types of stretching methods and the common methods are:

a. Jack stretching is to stretch and anchor prestressed reinforcements at the tops or the ends of the beams by jacks, which is especially suitable for curved reinforcements. It is always impractical for straight reinforcements because it is hard to put jacks at the end of a beam.

b. Transverse frapping is applying prestress across two directions. The principle is to fasten reinforcements at both ends, using simple tools including a torque-indicating wrench and blots to bend them from straight line and produce tensile strain; consequently, prestress is established in reinforcements.

The techniques of transverse frapping are as follows (shown in Fig. 3.12):

a) Fasten the ends of reinforcements ② to original beam. Reinforcements can be either curved lower-supported or straight line, as shown in Fig. 3.15(e).

b) Brace stay bars ④ (angle steel or thick steel bar) between two reinforcements ② at regular intervals.

c) Set U-shape screw ③ between stay bars and frap two reinforcements. Therefore prestress is established in reinforcements.

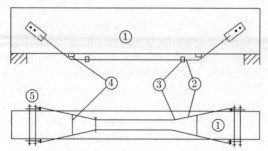

Fig. 3.12   Stretching prestress by transverse frapping method by manpower:
① original beam; ② retrofitted reinforcement; ③ U-shape screw; ④ brace stay bars;
⑤ high strength friction grip blot.

c. Vertical stretching includes stretching by manpower and by jacks.

The method by manpower is shown in Fig. 3.13. Fig. 3.13(a) shows vertical frapping; hooked frapping blot ③ is clawed to the reinforcements ② after drilling through the deep floor ④ (the initial shape of tie rod can be a straight line or curved line); when screwing the cap of frapping blot, reinforcements move downwards, which makes them become curved from straight or adds curvature; consequently, prestress is established. Fig. 3.13(b) shows vertical stretching by jacks, in which ⑦ is steel plate fixed at the bottom of beam, ⑧ is the lower-steel plate with nut welded on the reinforcements; when screwing the puller blot ⑥, the space between upper and lower steel plate is enlarged, reinforcements are forced to move down; consequently, prestress is established.

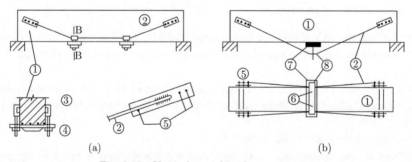

(a)   (b)

Fig. 3.13   Vertical stretching by manpower:
① original beam; ② retrofitted reinforcement; ③ frapping blot; ④ steel plate; ⑤ high strength friction
grip blot; ⑥ puller blot; ⑦ upper steel plate; ⑧ lower steel plate.

Fig. 3.14 shows prestressed reinforcements of roof truss by jacks. The techniques of retrofitting are:

a) Anchor two ends of reinforcements ① on plank.

b) Hang jacks ④ on the reinforcements by hooked stretch device ③ (oblique wedge is on the end of jack).

c) Start jacks, pull reinforcements apart from bearings ②, and insert steel pads into the gaps of reinforcements and bearings after stretching is qualified.

d. Electric heating stretching method. Apply large electric current at low electric pressure to the reinforcements, which can make them heat and elongate; cut off the current when the elongate quantity is qualified, and then anchor two ends immediately. After that, return material to room temperature and produce shrinkage distortion; consequently, prestress is established in reinforcements.

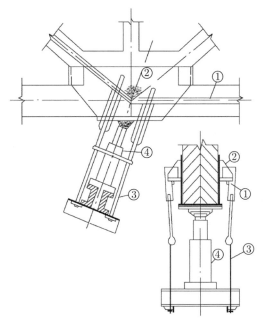

Fig. 3.14   Vertical stretching prestressed reinforcements by jacks:
① prestressed reinforcement; ② retrofitting bearing; ③ stretch device; ④ jack.

(2) Anchoring prestressed reinforcements
The common methods of anchoring prestressed reinforcements are below.
a. U-shape steel plate anchoring.
a) Chisel out concrete protective layer at the end of beam, and coat epoxy mortar on it.
b) Tightly fasten the U-shape steel plate which is as wide as beam on epoxy mortar.
c) Weld (end A in Fig. 3.15(a)) or anchor (end B in Fig. 3.15(a)) reinforcements to two sides of U-shape steel plate.
b. High strength friction grip bolt anchoring by friction and bonding according to the principle of high strength friction grip bolt in steel structure.

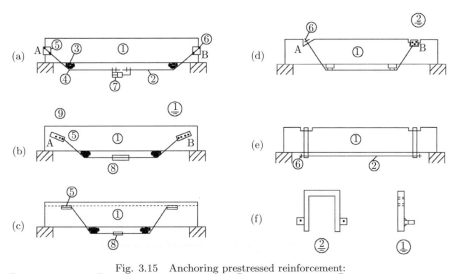

Fig. 3.15   Anchoring prestressed reinforcement:
① original beam; ② retrofitted reinforcement; ③ upper steel plate; ④ lower steel plate(bar);
⑤ welding; ⑥ screw; ⑦ external bracing jack; ⑧ anchoring joint; ⑨ high strength friction grip blot.

a) Drill holes with the same diameter as high strength friction grip bolt on the original beam and steel plate.

b) After coating epoxy mortar or high strength cement mortar on both steel plate and original beam, press steel plate tightly on original beam by high strength friction grip bolt, which can produce bonding and friction.

c) Anchor prestressed reinforcements on the flange which is welded with steel plate (end B in Fig. 3.15(b)) or weld prestressed reinforcements on steel plate directly (end A in Fig. 3.15(b)).

c. Welding and bonding anchoring involves welding reinforcements directly on the low stress region of original reinforcement and to felt by epoxy mortar (shown Fig. 3.15(c)). Low stress or even zero stress exists in certain sectors in steel bars in concrete beams (at inflection point of continuum beam or at the end of freely supported beam). This shows underutilization of steel strength and further potential. Consequently, reinforcements are welded on original steel bars of these sectors and bonded in the oblique groove by epoxy mortar.

d. Shoulder pole anchoring method adds steel plate (end A in Fig. 3.15(a)) or steel plate jacketing (end B in Fig. 3.15(d)) on the pressure region of original beam, and fix reinforcements on the steel plate (or jacketing). In construction, steel plate should be bonded on original beam by epoxy mortar to avoid slipping.

e. Anchoring by original preembedded piece. If there is an appropriate preembedded piece in the end of a retrofitted beam, welding reinforcement on this preembedded piece can achieve the anchoring aim.

f. Hooping anchoring is installing a jacket steel frame which is made of structural steel on the original beam, and anchoring prestressed reinforcements on steel modules. In construction, concrete protective layer of steel frame should be removed and steel frame should be fixed by epoxy mortar (shown in Fig. 3.15(e)).

2. Prestress retrofitting effect and internal force calculation

Prestressed reinforcements are in the external of retrofitted beams, and the force caused by them in original beam is usually contrary to the force caused by load (shown in Fig. 3.16), which creates 'unloading', so the flexions of retrofitted beam will be decreased, and the crack will be closed.

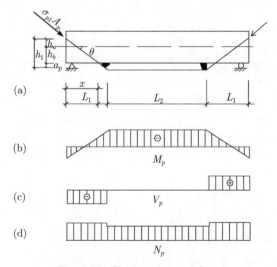

Fig. 3.16   Prestress internal force.

(1) Internal force analysis of retrofitted beam

The prestressed internal force, which is caused by lower-supported prestressed reinforcements in retrofitted beam, is shown in Fig. 3.16. The effective prestressed internal force in beam section L_1 is:

$$\left.\begin{aligned}
M_{px1} &= \sigma_{p1} \cdot A_p(h_a\theta - X\sin\theta) \\
V_{p1} &= \sigma_{p1} \cdot A_p \sin\theta \\
N_{p1} &= \sigma_{p1} \cdot A_p \cos\theta
\end{aligned}\right\} \tag{3.17}$$

The effective prestressed internal force caused by reinforcements in beam segment L_2 between two supporting points is:

$$\left.\begin{aligned}
M_{p2} &= \sigma_{p2} \cdot A_p(h_b + a_p) \\
V_p &= 0 \\
N_{p2} &= \sigma_{p2} \cdot A_p
\end{aligned}\right\} \tag{3.18}$$

where A_p is the total sectional area of prestressed reinforcements; σ_{p1}, σ_{p2} is the effective prestress of the prestressed reinforcement in beam segments L_1 and L_2, respectively, which is equal to the value that controls stress σ_{con} minus the loss of prestress force σ_l in each beam section (see detail case of σ_{con} and σ_l in the following text); X is the distance between anchoring point and calculation section; θ is the angle between oblique tensile bar and longitudinal axis; a_p is the distance between composite force of horizontal prestressed reinforcements and lower edge of section; h_a is the distance between anchoring point and longitudinal axis of original beam; h_b is the distance between longitudinal axis of original beam and lower edge of section.

The value of N_{p2} is a little less than the value of N_{p1} because of friction. When construction is finished, the value of internal force of section equals the difference between the internal force (M_0, V_0) caused by external loads and the internal force (M_p, V_p) caused by prestress, which is

$$\left.\begin{aligned}
M &= M_0 - M_p \\
V &= V_0 - V_p \\
N &= N_p
\end{aligned}\right\} \tag{3.19}$$

(2) Calculation of inverted camber and deflection of retrofitted beam

Prestress produces an invert arch in a retrofitted beam, so prestress retrofitting cannot only effectively strengthen beam, but also reduce deflection. When calculating retrofitted beam's deflection, the deflection f_1 before stretching, the inverted camber f_p caused by prestress and the deflection f_2 caused by later load after retrofitting should be considered, respectively, and then be superimposed together:

$$f = f_1 - f_p + f_2 \tag{3.20}$$

a. Calculation of f_1.

Undischarged load acts on the beam before stretching, which results in the deflection f_1. At this stage, the beam stiffness enhances a little with the increase of discharged load. However, due to the original beam long-term deformation, the stiffness under long-term load should be adopted when calculating deflection caused by undischarged load. Before the beam is retrofitted, beam stiffness, which is changing in a certain range, has related to the ratio of reinforcement and undischarged load and so on. For convenience, this stiffness is suggested to be:

$$B_1 = (0.35 \sim 0.5)E_cI_c \tag{3.21}$$

where E_c and I_c are elastic modulus of original concrete and inertia moment of transformed section, respectively.

For simply supported beam, f_1 can be approximately calculated in the following formula:

$$f_1 = \frac{5}{48}\alpha\frac{ML^2}{B_1} \tag{3.22}$$

where α is the influence coefficient of discharged load. The original beam has been unloaded, but the deflection cannot be recovered entirely because of concrete's plastic property and creep deformation etc., so α can usually be adopted as 1.1.

b. Calculation of f_p.

At the initial phase of stretching, owing to the existence of cracks at the bottom of beam, inverse stiffness is small and inverse deflection develops rapidly. With prestress increasing, cracks tend to close, stiffness enlarges and inverse deflection grows slowly. For simplification, stiffness is better considered as a constant. Considering that too large calculating value of inverted camber will influence members' safety, when calculating inverted camber using structural mechanics method, beam stiffness is suggested as the following formula:

$$B_p = 0.75E_cI_c \tag{3.23}$$

c. Calculation of f_2.

When retrofitting is finished, beam produces deflection f_2 under the later load. To calculate f_2, stiffness can be accounted as follows:

The beam with prestressed reinforcements in the external, retrofitted structure becomes a composite structure in fact, in which tensile bar is prestressed reinforcement, while the original beam works as an arch. The deflection can be calculated with structural mechanics. For simplification, the method above-mentioned can also be used to calculate this deflection, and retrofitted beam stiffness is suggested to be calculated by the following formula:

$$B_2 = (0.7 \sim 0.8)E_cI_c \tag{3.24}$$

3. Bearing capacity calculation of retrofitted beam

(1) Bearing capacity calculation of normal section

As to retrofitted beam with prestressed reinforcements after being retrofitted, prestressed reinforcements contact with original beam only at anchoring point and supporting point. When the beam deflects with the increasing of loads, original reinforcement in beam elongates with the increasing of curvature of original beam; however, deformation of prestressed reinforcements is not the same as that of original reinforcements, and relates only to the beam deflection at supporting point and anchoring point. Certain research indicates that stress increment ratio in prestressed reinforcements is far less than that in original reinforcement (only 18%~35%). Because of this deformation incongruity, equivalent load method is used to calculate sectional bearing capacity of this kind of retrofitting beam.

Equivalent load means that action by prestress in original beam can be replaced by corresponding load, and the internal forces (moment, shear force and axial force) of original beam are equal. Equivalent load method is to apply prestress as equivalent load on original beam in calculating bearing capacity of retrofitted beam, and then verify bearing capacity of original beam according to its size and distributed steel condition.

The stress of prestressed reinforcements is the sum of the stress when stretching is finished and the stress increment that is caused by later load. However, this kind of stress increment is actually very small, and is almost the same value (about 5.35 MPa) per increasing unit load, which exerts little influence on the stress of prestressed reinforcements. When high strength prestressed reinforcements are adopted, areas of reinforcements become smaller, so this influence to the total prestress internal force becomes even less. For convenient calculation, this stress increment can be neglected in design, and internal force of prestressed reinforcements is adopted as σ_{con}, which is a little higher than the stress value after stretching, but is inclined to be safe.

The original bending member becomes an eccentric compression member, usually a very eccentric compression member because of longitude prestress N_p. Therefore, bearing capacity of retrofitted beams can be verified by the formula of large eccentric compression in current national standards of the code for design of concrete structure, shown in Fig. 3.17.

$$N \leqslant f_{cm}bx + f_y'A_s' - f_yA_s$$
$$Ne \leqslant f_{cm}bx\left(h_0 - \frac{x}{2}\right) + f_y'A_s'(h_0 - a_s') \tag{3.25}$$

f_{cm} is the design value of flexural compressive strength of concrete, and can be used as $1.1f_c$; A_s and A_s' are sectional area (with no consideration of negative constructional steel in beam) of tensile steel and compression steel in original beam, respectively; x is the height of concrete compressive region; f_y and f_y' are strength design value of tensile steel and compression steel in original beam, respectively; e is the distance from the action spot of longitude force N to the gravity center of tensile steel A_s, which is

$$e = \eta e_i + \frac{h}{2} - a_s \tag{3.26}$$

e_i is eccentric distance of longitude force N, $e_i = e_0 + e_a$; e_a is extra eccentric distance of longitude force; e_0 is the distance from the action spot of longitude force N to sectional gravity center; η is the enhancement coefficient of eccentric distance; M and N are moment and longitude force acting on the section, respectively, which are calculated as Eq. (3.19).

Moreover, according to recent research, the bearing capacity of retrofitted beams with exposed reinforcements can be calculated as beams with unbonded tendons, and the strength of retrofitted reinforcements can be adopted as $0.8f_{py}$.

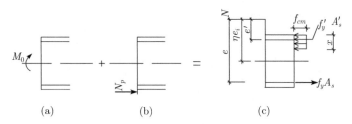

(a)   (b)   (c)

Fig. 3.17   Sectional internal force of beam with prestressed external reinforcement.

(2) Bearing capacity calculation of oblique section

As mentioned above, mechanical characteristics of a retrofitted beam with exposed prestressed reinforcements is the same as eccentric compressive components, so compared with original beam, shearing resistance of retrofitted beam is enhanced. To the beam retrofitted by straight prestressed reinforcements, the enhancement is dependent on longitude force N_p, which is caused by prestress. Consequently, the oblique sectional shearing resistance of this beam should be the sum of original beam's shearing resistance and the shearing resistance enhancement caused by longitude force N_p, which is

$$V \leqslant V_u' + 0.05N_p \tag{3.27}$$

For curved prestressed reinforcements, the shearing resistance of retrofitted beam is:

$$V \leqslant V_u' + (\sigma_{con} - \sigma_l)A_p\sin\theta + 0.05N_p \tag{3.28}$$

where V_u' is oblique sectional shearing resistance of original beam; σ_l is the total loss of prestress.

4. Stretching value calculation

In retrofitting engineering, stretching value of prestressed reinforcements is usually used to control stretching stress. The methods of stretching transversely by manpower and electric heating are adopted in many retrofitting projects. However, the stretching value calculation of retrofitted reinforcements is more complex. This is not only because of variety kinds of stretching methods for retrofitted reinforcements and multiple unstable factors in construction, but also because of the complex environment of retrofitted engineering. For example, during the process of stretching prestress, it is inevitable for the cracks of the original beam to close. This phenomenon influences stretching value a lot. Formulas, which are used to calculate stretching value of prestressed reinforcements in retrofitted beam, are introduced in the following:

(1) The calculation of shrinkage distortion induced by crack closing

When tensioning the prestressed reinforcements, crack closure of original beam will result in shrinkage, which will increase stretching value of retrofitted reinforcement. Cracks of retrofitted beam are usually wide, and closure distortion caused by stretching prestress is also large. This kind of shrinkage distortion sometimes influences prestress effect a lot. For example, a certain beam, spanning 6 m, needs to be retrofitted because of too wide of a crack. There are 10 cracks on a beam with average width of 0.3 mm before being retrofitted. Grade II steel bar is used to retrofit with prestressed reinforcement's length of 5 m. Prestress value of 126 MPa will be lost (50% of stretching control stress) if no shrinkage distortion is considered in the calculation. This shows that shrinkage distortion influences stretching a lot. Always, actual value of stretching is up to the calculation value; however, the actual value of prestress in retrofitted reinforcements is relatively small. Furthermore, considering the prestress loss exists in prestressed reinforcements, the effect of prestress is even worse. So crack shrinkage distortion attaches much influence to stretching value of prestressed reinforcements, which should be considered in retrofitting projects.

The original beam's crack closure distortion caused by prestress is the difference of distortion in original beam's reinforcements before and after tension. The distortion of original beam's reinforcements equals the average strain of original reinforcements multiplied by beam's length. If the original beam is still used during prestress period, formulas provided in codes can be used to deduce the following formulas for calculating original beam's total distortion ΔS_i in the action of the i load:

$$\Delta S_i = \varepsilon_{si} \cdot L = \phi \frac{\sigma_{si}}{E_s} L \tag{3.29}$$

where L is horizontal length of prestressed reinforcement; σ_{si} is the stress of original tensile reinforcement at crack section; calculation formula of σ_{si} is:

$$\sigma_{si} = \frac{M_i}{0.87 h_0 A_s} \tag{3.30a}$$

where M_i is the load moment of members section under No. i load; A_s is the section area of original tensile reinforcements; ϕ is the inhomogeneous coefficient of steel strain between cracks.

$$\phi = 1.1 - \frac{0.65 f_{tk}}{\rho_{et} \sigma_{si}}, \quad 0.4 \leqslant \phi \leqslant 1.0 \tag{3.30b}$$

where f_{tk} is the axial tensile characferistic strength of concrete; ρ_{et} is the effective ratio of reinforcement of original beam's tensile reinforcement. $\rho_{et} = \dfrac{A_s}{A_{et}}$, A_{et} is the effective tensile area of concrete, which is usually adopted as the section area under central axis. As to rectangle section,

$$\rho_{et} = \frac{A_s}{0.5bh}$$

where h, h_0 is sectional height, sectional effective height of original beam, respectively.

The particular calculation steps of shrinkage distortion, which occur when the beam is stretched, are as follows:

a. Based on M_1, which is the moment acted on the beam before prestress is applied, extension value ΔS_i of original reinforcements before retrofit can be calculated by Eq. (3.29).

b. Compute internal force caused by external prestressed reinforcements. The method is to solve internal force, considering prestress as external force to apply on original beam.

c. Solve the action of moment M_2 after original beam is applied prestress, which is the difference between load moment M_1 and prestressed moment M_p, that is $M_2 = M_1 - M_p$.

d. According to M_2, judge the beam stress after being retrofitted, and calculate extension value of original reinforcements.

If $M_2 > M_{cr}$, crack has not been closed, and the remnant extension ΔS_2 of original reinforcement can be calculated as Eq. (3.29).

If $0 \leqslant M_2 \leqslant M_{cr}$, crack is almost closed, or the remnant crack's width is very small. For convenience, it can be adopted as ΔS_2.

If $M_2 < 0$, crack is already closed, original reinforcements have changed from tensile status to compression status, so $\Delta S_2 = 0$. Where, M_{cr} is crack moment of original beam, and can be calculated as formula in codes. As to rectangle section,

$$M_{cr} = 0.235bh^2 f_{tk}$$

e. Calculate shrinkage distortion ΔS of beam

$$\Delta S = \Delta S_1 - \Delta S_2 \tag{3.31}$$

(2) Stretching value calculation when using jack to stretch

a. Stretching value calculation of straight-line reinforcements. As to prestress straight reinforcements, stretching value ΔL can be obtained by the following formula:

$$\Delta L = \frac{\sigma_{con}}{E_s}L + 2a + \Delta S \tag{3.32}$$

where L is the distance between stretching point and anchoring point; α is anchor device deformation at anchoring point and prestress reinforcements' shrinkage value (Table 3.5); ΔS is beam's longitude close shrinkage distortion caused by prestress (calculation by Eq. (3.31)); σ_{con} is stretching control stress of prestressed reinforcements.

Anchor device deformation should be considered if stretching is performed in the middle regions of retrofitted reinforcements.

b. Stretching value calculation of curved reinforcements.

When curved reinforcements are stretched by jacks, the method of stretching at two ends is usually used (shown in Fig. 3.18). There is more prestress loss caused by friction at lower supporting points. When one end is stretched, a lot of stress is lost at the other end. Applying two-end stretching, total stretching value ΔL can be calculated as the following formula:

$$\Delta L = \frac{\sigma_{con} - \sigma_{l2}}{E_s}L_2 + 2\frac{\sigma_{con}}{E_s}\sqrt{L_1^2 + h_1^2} + \Delta S \tag{3.33}$$

where σ_{l2} is prestress loss caused by friction at lower-supporting point; L_1 is horizontal length of oblique bar; L_2 is the length of horizontal reinforcements between two supporting points; h_1 is the height of prestress oblique reinforcement.

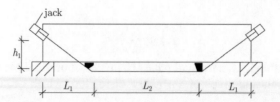

Fig. 3.18   Two-end stretching curved reinforcement.

When jack is put at the middle region of beam, stretching value ΔL should be calculated by the following formula:

$$\Delta L = \frac{\sigma_{con}}{E_s}L_2 + 2\frac{\sigma_{con} - \sigma_{l2}}{E_s}\sqrt{L_1^2 + h_1^2} + 2a + \Delta S \tag{3.34}$$

(3) Stretching value calculation when transversely frapping.

The technique is to frap retrofitted reinforcements across directions, and elongate them lengthwise, at the condition that two ends are both anchored. This method is relatively applicable for beams with large sections, because when section is small, frapping amount is small, and high prestress is difficult to establish. If beam section is wide, and span is not very large, one-point frapping can be used; otherwise, two-point frapping or multi-point frapping should be adopted. However, there cannot be too many frapping points to reduce inhomogeneous stress of prestressed reinforcements along the longitude direction.

Now let's focus on the function relationship between transversely frapping value ΔH and longitude extension value ΔL of both straight and curved reinforcements.

a. Frapping straight reinforcements at two points.

Fig. 3.19 shows a retrofitted beam with straight prestressed reinforcements by two supporting rods frapped at two points. Prestressed reinforcement ① before being trapped is positioned at abcdefg, corresponding reinforcement ② at mn. Set supporting bar at point b and point f, and then frap two bars by U-shape screw at point c and point e. After being frapped, reinforcement ① is moved to position $a'b'c'd'e'f'g'$, and reinforcement ② is also moved correspondingly. So the length between point b' and c' is $L_1 + \Delta L_1$, horizontal length is $L_1 - \Delta L_3 - (\Delta L_L/2)$. By geometrical relationship, obtain

$$\Delta H^2 = (L_1 + \Delta L_1)^2 - \left(L_1 - \Delta L_3 - \frac{\Delta S_L}{2}\right)^2$$

To develop the above formula, neglect higher order micro contents of $\Delta L_1^2, \Delta L_3^2, \Delta S_2^2$ and $\Delta L_3 \times \Delta S_L$, get

$$\Delta H^2 = L_1(2\Delta L_1 + 2\Delta L_3 + \Delta S_L)$$

Take $2L_1 + 2L_3 = L_0, 2\Delta L_1 + 2\Delta L_3 = \Delta L_0$, then the transversely frapping value ΔH is

$$\Delta H = \sqrt{L_1(\Delta L_0 + \Delta S_L)} \tag{3.35}$$

where L_1 is the distance from horizontal supporting bar to its adjacent frapping point (Fig. 3.19); ΔL_0 is stretching extension value of prestressed reinforcement at the region of L_0, which can be calculated as the following formula:

$$\Delta L_0 = \frac{\sigma_{con}}{E_s}L_0 \tag{3.36a}$$

ΔS_L is the sum of closure shrinkage deformation, extension value of straight prestressed reinforcements and deformation of anchor device, which can be obtained as:

$$\Delta S_L = \Delta S + \frac{\sigma_{con}}{E_s} \cdot 2L_2 + 2a \tag{3.36b}$$

where L_2 is the length of prestressed reinforcements from anchoring point to the first supporting bar (Fig. 3.19); ΔS is the closure shrinkage deformation of original beam between two prestressed reinforcements' anchoring points (as Eq. (3.31)); a is the deformation of anchor device, and $a = 0$ when point a and g are welding points.

b. Frapping straight reinforcements at one point.

Fig. 3.20 shows a straight prestressed reinforcement that is frapped at one point. After being frapped, prestressed reinforcement is moved from position abcde to position $a'b'c'd'e'$. Compared to Fig. 3.19, when $L_2 = 0$, two-point frapping becomes one-point frapping, so one-point shrinkage value can be calculated as Eq. (3.36), where $L_2 = 0$ is ordered.

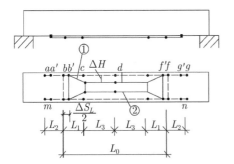

Fig. 3.19   Frapping prestressed reinforcement by two supporting rods at two points.

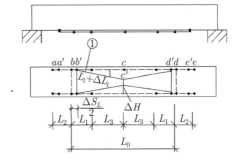

Fig. 3.20   Frapping prestressed reinforcement by two supporting rods at one point.

c. Transversely frapping curved reinforcement.

When curved reinforcement is frapped transversely, the shrinkage value ΔH can still be calculated as Eq. (3.35); however, ΔS of ΔS_L represents original beam's closure shrinkage distortion at the range of curved reinforcement's horizontal length L' which is under the natural axis. That is to say, L should be given the horizontal length of curved reinforcement under the natural axis when ΔS is calculated. As to one-point frapping, $L_3 = 0$ in Eq. (3.36).

(4) Vertical stretching value calculation

Vertical stretching prestressed reinforcements cannot only produce opposite moment in original beam as all kinds of prestress methods mentioned in the above, but can also produce opposite load at original beam's top bracing point, resulting in larger opposite deflection than the other prestress methods. This opposite deflection will counteract some vertically stretching value, and decrease prestress effect. Consequently, this disadvantage factor of opposite deflection should be considered in deducing formula of vertically stretching value.

When anchoring point of retrofitted reinforcement is near neutral axis, closure shrinkage distortion will not cause an anchoring point to move; when the anchoring point is above the natural axis, the anchoring point will move outwards. Ordinarily, anchoring point of retrofitted reinforcement applying prestress by vertical stretching lays near or above the neutral axis. So closure shrinkage distortion is not considered in the formula of vertically stretching value calculation.

The initial position of prestressed reinforcement can be classified into straight line and curved line. The following text takes curved prestressed reinforcement that is stretched at two points for example, to illustrate calculation formula of stretching value.

a. Two-point vertically stretching curved reinforcement.

Fig. 3.21 shows vertically stretching curved prestressed reinforcement at two points to retrofit beam. Before being stretched, prestressed reinforcement is positioned at abcd. After point b and point c are vertically stretched, point b moves to b', while point c moves to

c', and L_1 elongated to $L_1 + \Delta L_1$, horizontal length of which is $L_1' - \Delta L_1$. According to geometrical relationship and original beam's opposite deflection caused by prestress,

$$(\Delta H + H - f)^2 = (L_1 + \Delta L_1)^2 - (L_1' - \Delta L_2)^2$$

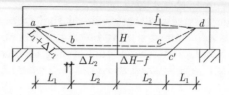

Fig. 3.21 Stretching curved prestressed reinforcement retrofitted beam at two points.

Develop the right item in the above formula, neglect higher order micro contents of $\Delta L_1^2, \Delta L_2^2$, and as $L_1^2 - L_1'^2 = H^2$, then get

$$(\Delta H + H - f)^2 = H^2 + 2L_1 \Delta L_1 + 2L_1' \Delta L_2$$

So the vertical top bracing value ΔH can be obtained by the formula:

$$\Delta H = f - H + \sqrt{H^2 + 2(L_1 \Delta L_1 + L_1' \Delta L_2)} \tag{3.37}$$

where L_1 and L_1' are oblique reinforcement's initial length and its horizontal length, respectively; ΔL_1 is retrofitted reinforcement's deformation at the segment L_1:

$$\Delta L_1 = \frac{\sigma_{con}}{E_s} L_1 + a$$

ΔL_2 is retrofitted reinforcement's deformation at the segment L_2 (horizontal segment/2); H is initial kink length of prestressed reinforcement (Fig. 3.21); f is opposite deflection caused by prestress internal force. When opposite deflection is calculated, beam stiffness is computed as Eq. (3.23). The factor that f influences on ΔH can be neglected if prestress is small.

b. One-point vertical stretching curved reinforcement.

Fig. 3.22 shows vertical stretching curved prestressed reinforcement at one point to retrofit beam. The difference from two-point stretching method lies on $L_2 = 0, \Delta L_2 = 0$. Consequently, calculation formula of stretching value can be obtained by substituting $\Delta L_2 = 0$ into Eq. (3.37), and ordering $\Delta L = 2\Delta L_1$, which is

$$\Delta H = f - H + \sqrt{H^2 + L_1 \Delta L_1} \tag{3.38}$$

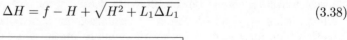
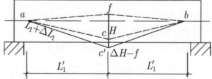

Fig. 3.22 Stretching curved prestressed reinforcement retrofitted beam at one point.

c. Vertical stretching straight-line reinforcement.

Fig. 3.23 shows stretching straight line prestressed reinforcement by one point and two points method to retrofit beam. The difference from curved prestressed reinforcement is $L_1 = L_1', H = 0$. Substituting them into Eq. (3.37) and adopting L as the whole length of tensile bar, can determine uniform expression of straight line reinforcement by one-point stretching and two-point stretching method, which is

$$\Delta H = f + \sqrt{L_1 \Delta L} \tag{3.39}$$

where L_1 is the length of prestressed reinforcement from anchoring point to stretching point; ΔL is the distortion of whole prestressed reinforcement.

$$\Delta L = \frac{\sigma_{con}}{E_s} L + 2a$$

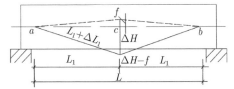 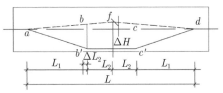

Fig. 3.23   Stretching straight prestressed reinforcement at one point and at two points to retrofitting beam.

(5) Stretching value calculation when using electric heating method to stretch

The formula, which is to calculate stretching value ΔL of stretching prestressed reinforcement by electric heating method, is:

$$\Delta L = \frac{\sigma_{con} + 30}{E_s} L + \Delta S + 2a \tag{3.40}$$

where L is the length of prestressed reinforcement (mm), 30 is additional prestress loss caused by nonstraight prestressed reinforcement and its plastic deformation in high temperature.

The power of transformer that is needed to elongate tensile bar by ΔL is

$$P = \frac{G \cdot C \cdot t}{1.59T} \ (\text{kV} \cdot \text{A}) \tag{3.41a}$$

where G is steel bar's weight (kg) which is stretched at the same time, C is steel bar's thermal capacity, given 0.481×10^{-3} J/(kg $\cdot$ K); T is steel bar's electric heating duration (h); t is the needed temperature to elongate steel bar.

$$t = t_0 + \Delta t$$

$$\Delta t = \frac{\Delta L}{\alpha \cdot L} \tag{3.41b}$$

t_0 is environment temperature when stretching by electric heating method, Δt is increased temperature when stretching by electric heating method and α is linear expansion factor of steel reinforcement, commonly 0.000012.

5. Stretching control stress and prestress loss

Stretching control stress and prestress loss should be reasonably determined in order to comprehend retrofitted beam's stress and its change before and after stretching, and to control retrofitted beam's performance and effect during stretching and retrofitting period.

(1) The value of stretching control stress

Tensile steel reinforcement stress in retrofitted beam is usually high, so, the difference between stress of prestressed reinforcement and stress of original beam's tensile reinforcement in retrofitting beam is much less than the difference between these two kinds of stress in an ordinary prestressed concrete beam. In addition, deflection of beam to be retrofitted is relatively large, and so is the crack width. This shows that the higher the value of prestress that is applied on retrofitted reinforcement, the more the stress condition of retrofitted beam can be improved. Consequently, stretching control stress σ_{con} should not be too high, or the internal force of some steel bars will arrive at or exceed the yield strength and result in danger during over-stretching period. Stretching control stress is listed in Table 3.4.

Table 3.4   Allowed stretching control stress σ_{con}

Item number	Steel category	Stretching control stress σ_{con}
1	Carbon wire, indented wire, steel stranding wire	$0.70f_{ptk}$
2	Cold-drawn low-carbon wire, heat treated steel bar	$0.65f_{ptk}$
3	Cold-drawn hot-rolled steel bar	$0.85f_{pyk}$
4	Hot-rolled steel bar	$0.90f_{pyk}$

As retrofitted reinforcement produces small prestress to original beam's concrete, the concrete creep loss is accordingly small, or even does not exist. That is to say, under the same condition of stretching control stress, the ultimate stress of retrofitted reinforcement is higher than that of prestressed reinforcement.

(2) Calculation of prestress loss

Constitutions and techniques of external prestressed reinforcement of retrofitted beams are both different from ordinary prestressed concrete beams, and so is prestress loss. For convenience, the following text will make the most of signs regulated by *Code for Design of Concrete Structure*.

a. Anchoring loss σ_{l1}. σ_{l1} can be calculated as the following formula:

$$\sigma_{l1} = \frac{a}{L} E_s \tag{3.42}$$

where L is effective length of prestressed reinforcement when straight retrofitted reinforcement is stretched at the end or in the middle by jack; and is half of effective length of prestressed reinforcement when curved reinforcement is stretched at the end (often both two ends are stretched simultaneously). a is anchor device distortion value and steel reinforcement retraction value. Values in *Code for Design of Concrete Structure* are shown in Table 3.5. In retrofitting, the anchor device is much more complex than that in ordinary prestress concrete beams. When anchor device is not included in Table 3.5, a can be determined by referencing Table 3.5 and adapting to actual situation.

Table 3.5   Anchor device distortion and steel reinforcement retraction values

Item number	Anchor device distortion	Steel reinforcement retraction value (mm)
1	Anchor device with screw cap (including cone screw anchor device, cylinder anchor device, etc.)	
	Gap between screw caps	1
	Gap between additional pads	1
2	Heading anchor device of steel tendon	1
3	Steel cone anchor device of steel tendon	
	JM-12 anchor device	5
4	When prestressed reinforcement is steel bar	3
	When prestressed reinforcement is steel stranding wire	5
5	Cone anchor device of single cold-drawn low-carbon wire	5

Distortion loss caused by bar stay or screw can be neglected. These kinds of distortion exist in stretching, and prestress value is constant before and after anchoring.

b. Prestress loss of friction in turning point σ_{l2}. When down-supported prestressed reinforcement is applied, friction at lower-supporting point will make internal force of retrofitted reinforcement at the other point smaller than that at stretching end. That is to say, pre-stress will be reduced because of friction at turning point. Prestress loss of friction σ_{l2} can be calculated by formula in *Code for Design of Concrete Structure*.

$$\sigma_{l2} = \sigma_{con}(1 - e^{-\mu\theta}) \tag{3.43}$$

where θ is the included angle from oblique retrofitted reinforcement to longitudinal axis; μ is friction coefficient between supporting pad and gliding block. When they are both steel plate, $\mu = 0.25$; when retrofitted reinforcement contacts steel gliding plate directly, $\mu = 0.4$; and when jacket is placed in supporting point, $\mu = 0.1$.

c.   Prestress loss of steel relaxation loss σ_{l4}.   Prestress loss caused by relaxation of retrofitted reinforcement can be adopted as value that is regulated in *Code for Design of Concrete Structure*.   Cold-drawn hot-rolled steel bar and heat treated steel bar could be calculated as the following formula:

$$\text{One-time stretching, } \sigma_{l4} = 0.05\sigma_{con} \tag{3.44}$$

$$\text{Over-stretching, } \sigma_{l4} = 0.035\sigma_{con} \tag{3.45}$$

Prestress loss of relaxation loss of steel wire and steel stranding wire can be calculated as the following formula:

$$\sigma_{l4} = \psi\left(0.36\frac{\sigma_{con}}{f_{pyk}} - 0.18\right)\sigma_{con} \tag{3.46}$$

where ψ is empirical coefficient related to stretching technique: One-time stretching, $\psi = 1.1$; over-stretching, $\psi = 0.9$; f_{pyk} is standard strength of prestressed reinforcement.

d. Prestress loss of concrete creep loss σ_{l5}. Because prestressed reinforcement in retrofitted member is only one part of tensile steels (or even small part), and it produces very small compressive pre-stress in concrete, which is not large enough to counteract the tensile stress caused by external load which is acted at the same time with prestress, concrete creep loss σ_{l5} can be neglected.

6. Constructional detail requirements

Concrete slab and beam structure retrofitted by prestress should follow constructional details as below.

a. It is better to adopt steel reinforcement bar or steel stranding tendon with diameter $\phi12 \sim \phi30$; when choosing prestressed reinforcement, diameter is better as $\phi4 \sim \phi8$.

b. When slab is retrofitted by prestress, flexile steel wire, not thick steel bar, is preferred.

c. Clear distance from horizontal segment of straight prestressed reinforcement or lower-supported prestressed reinforcement to bottom of retrofitted beam is better to be: <100 mm, and 30~80 mm is more suitable.

d.   External retrofitted reinforcement needs anti-corrosive treatment after stretching, which includes gunning cement mortar method and brushing anticorrosion paint method.

e. When stretching transversely method is used, diameter of frapping screw is better to be: $\geqslant \phi16$, and nut height should be no less than 1.5 times screw diameter.

f. Anchoring of prestressed reinforcement should be firm and reliable, and without dis-placement.

g. Supporting steel pad should be set on the bottom of original beam, at the turning point of lower-supported prestressed reinforcement. The height of steel pad should be $\geqslant 10$ mm, the width should be no less than 4 times its height, and the length should be as long as the width of retrofitted beam.

Steel bar (shown in Fig. 3.24(a)), of which diameter $\geqslant 20$ mm, length not less than $b + 2d + 40$ (b is beam width, d is diameter of prestressed reinforcement) or steel plate (shown in Fig. 3.18) should be set between supporting steel pad and prestressed reinforcement. Sometimes steel cylinder, of which length equal to the width of beam, is enveloped out of steel bar to reduce friction prestress loss (shown in Fig. 3.24(b)).

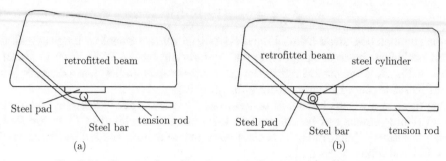

Fig. 3.24   Construction of turning point of prestressed reinforcement.

When a continuous slab is retrofitted by prestress, the turning point of prestressed reinforcement is better to be positioned around inflection point, which can produce obvious upward force and reduce the span of slab (Fig. 3.25).

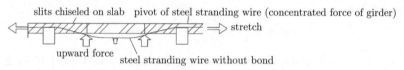

Fig. 3.25   Retrofitting continuous slab by prestress method.

h. Oblique hole through which reinforcement goes at turning point of prestressed reinforcement of continuous slab can be at 45°, and its position should avoid reinforcements in slab. From oblique hole, slits should be chiseled on slab face and bottom, respectively, along direction of prestressed reinforcement. The depth of slits depends on the requirements of upward force at turning point: the shallower the slits, the larger the upward force, and more prestress loss at turning point.

i. Two-end stretching is better to stretch prestressed reinforcement of continuous slab, in order to reduce friction prestress loss.

7. Retrofitting design examples

(1) Calculation steps
For the retrofitted beams with prestressing reinforcements in external, calculation steps can be generalized as follows:

a. Draw internal force map under residual load and whole load.

b. First check the value of compression region height, x, considering the member as a flexural member with bending and then, moment M, obtain the moment ΔM resulting from retrofitting reinforcement in the mid-span section of the beam, finally estimate the cross-sectional area of retrofitting reinforcement Ap, according to ΔM.

c. Determine tension control stress, and calculate the loss of prestress.

d. Based on the tension control stress, calculate prestress internal forces.

e. Act prestress as external force on the original beam, check the normal section bearing capacity of the original beam considering it as an eccentric compressive members. If the results cannot meet the demands, increase the area of prestressing reinforcements and recheck it.

f. Check the oblique-sectional bearing capacity of the beam.

g. Proceed calculation of the prestressing effect and extension value.

(2) Calculation examples

Example 3.3   A crossbeam of an equipment platform in a steel factory with a calculative span of 9 m, bears a uniformly distributed load of 19.7 kN/m, uniform live load of 14 kN/m, and equipment load of 26 kN/m at mid-span (see Fig. 3.26). Now the equipment needs to be replaced, so the load in mid-span is increased to 96 kN. Try to retrofit this beam (all of the loads above are design loads).

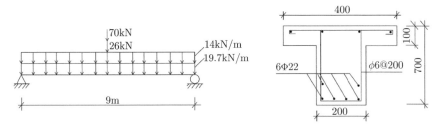

Fig. 3.26   Load and section of equipment crossbeam.

Solution:

a.   Conditions of the original beam.   The tensile main bar is 6Φ22, $A_s = 2281$ mm^2, $f_y = 310$ N/mm^2, the strength grade of concrete is C20, $f_{cm} = 13.5$ kN/m, $h_0 = 700 - 60 = 640$ mm, the distance between the centroid of section and lower edge $y_0 = 442.9$ mm, the inertia moment of section $I_0 = 1.0 \times 10^{10}$ mm^4.

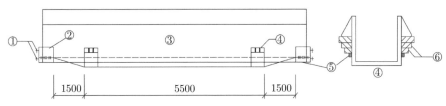

Fig. 3.27   Retrofitting schematic of crossbeam.
① expansion bolts; ② U-shaped anchoring plate; ③ original beam; ④ U-shaped buttress plate;
⑤ prestressed retrofitting reinforcement; ⑥ steel pad.

b.   Retrofitting method.   Adopting vertical top-supporting method (Fig. 3.27), two ends of the prestressed reinforcements are anchored by U-shaped plates. In order to prevent the U-shaped plates from slipping down, four expansion bolts are used at the ends of U-shaped plates to make them fixed. The method is vertical stretching by jacks. When the prestressed reinforcements are stretched to the right place, steel plates should be set between anchoring point and prestressed reinforcements with spot welding, thus the prestressed reinforcements work outside the beam.

c.   Calculate the internal forces.   While being retrofitted, there is only distributed dead load on the beam, and the bending moment is

$$M_0 = \frac{1}{8} \times 19.7 \times 9^2 = 199.46 \text{ kN} \cdot \text{m}$$

Under whole load

$$M_{\max} = \frac{1}{8}(19.7 + 14) \times 9^2 + \frac{1}{4} \times 96 \times 9 = 557.2 \text{ kN} \cdot \text{m}$$

$$V_{\max} = \frac{1}{2}(19.7 + 14) \times 9 + \frac{1}{2} \times 96 = 199.65 \text{ kN} \cdot \text{m}$$

d. Obtain the moment ΔM resisted by prestressed reinforcements. Cold-drawn grade III steel bars are chosen as reinforcing tie rods

$$f_{py} = 420 \text{ N/mm}^2$$

Judge the type of T-shaped beam

$$f_{cm} b'_f h'_f \left(h_{01} - \frac{h'_f}{2} \right) = 13.5 \times 400 \times 100 \times \left(665 - \frac{100}{2} \right)$$

$$= 3.321 \times 10^8 \text{ N} \cdot \text{mm} < M = 5.572 \times 10^8 \text{ N} \cdot \text{mm}$$

This result shows that the beam belongs to the second T-shaped beam type.
Obtain the height of compression region, x,

$$x = h_{01} \left(1 - \sqrt{1 - \frac{2M}{f_{cm} b h_{01}^2}} \right)$$

$$= 665 \left(1 - \sqrt{1 - \frac{2 \left[5.572 \times 10^8 - (400 - 200) \times 100 \times 13.5 \left(665 - \frac{100}{2} \right) \right]}{13.5 \times 200 \times 665^2}} \right)$$

$$= 665(1 - 0.587) = 274.7 \text{ mm}$$

Obtain the distance between centroid of the section and upper edge,

$$y'_0 = \frac{(400 - 200) \times 100 \times \dfrac{100}{2} + 200 \times 274.7 \times \dfrac{274.7}{2}}{(400 - 200) \times 100 + 200 \times 274.7} = 114 \text{ mm}$$

Thus,

$$\Delta M = M_{\max} - A_s f_y (h_0 - y'_0)$$

$$= 5.572 \times 10^8 - 2281 \times 310 \times (640 - 114) = 1.853 \times 10^8 \text{ N} \cdot \text{mm}$$

e. Estimate the sectional area of prestressed reinforcements.

$$A_p = \frac{\Delta M}{f_{py}(h + a_p - y'_0)} = \frac{1.853 \times 10^8}{420 \times (700 + 15 - 114)} = 740 \text{ mm}^2$$

Choose 2Φ22, $A_p = 760 \text{ mm}^2$.
f. Determine the tension control stress, and calculate loss of prestress. According to Table 3.4, the control stress may be obtained as

$$\sigma_{con} = 0.85 f_{pyk} = 0.85 \times 500 = 425 \text{ N/mm}^2$$

Calculation of prestress loss, because the prestressed reinforcements are anchored by welding,

$$\sigma_{l1} = 0$$

The initial state of the prestressed reinforcements is linear, the vertical stretching value is small, so

$$\sigma_{l2} = 0$$

The loss of prestress caused by stress relaxation is

$$\sigma_{l4} = 0.05 \sigma_{con} = 21 \text{ MPa}$$

g.   Calculate the internal forces caused by prestress and its effect. As to the beams retrofitted by exposed prestressed reinforcements, the section strength should be checked considering the beam as eccentric compressive member.

h. Based on σ_{con}, calculate the prestress internal forces

$$N_p = \sigma_{con} A_p = 425 \times 760 = 3.23 \times 10^5 \text{ N}$$

$$M_p = N_p(y_0 + a_p) = 3.23 \times 10^5 \times (442.9 + 15)$$
$$= 1.48 \times 10^8 \text{ N} \cdot \text{mm} = 148 \text{ kN} \cdot \text{m}$$

i. Check the normal section bearing capacity of original beam, considering it as an eccentric compressive member.

External bending moment M acted on the beam is

$$M = M_{\max} - M_p = 557.2 - 148 = 409.2 \text{ kN} \cdot \text{m}$$

$$e_0 = \frac{M}{N} = \frac{409.2 \times 10^5}{3.23 \times 10^5} = 1267 \text{ mm} > 0.3 \, h_0$$

Then

$$e_a = 0, \quad e_i = e_0 + e_a = 1267 \text{ mm}$$

Assume $\eta = 1$, $\eta e_i = 1267 > 0.3 \, h_0$, check it as large eccentric compression member firstly. Tentatively calculate the value of x (taking no account of the compressive constructional reinforcements) according to Eq. (3.25)

$$x = \frac{N_p + f_y A_s}{f_{cm} b_i'} = \frac{3.23 \times 10^5 + 2281 \times 310}{13.5 \times 400}$$
$$= \frac{1.03 \times 10^6}{5400} = 190 \text{ mm} > h_i' = 100 \text{ mm}$$

This shows that neutral axis goes through the flange of the section. x needs to be recalculated:

$$x = \frac{1.03 \times 10^6 - f_{cm}(b_i' - b)h_i'}{f_{cm} b} = \frac{1.03 \times 10^6 - 13.5 \times (400 - 200) \times 100}{13.5 \times 200}$$
$$= 280 \text{ mm} < \xi_b h_0 = 0.53 \times 640 = 339 \text{ mm}$$

(belongs to large eccentric compression members).

$$e = \eta e_i + y_0 - a_s = 1267 + 442.9 - 60 = 1650 \text{ mm}$$

$$Ne = 3.23 \times 10^5 \times 1650 = 5.33 \times 10^8 \text{ N} \cdot \text{mm}$$

$$f_{cm} bx \left(h_0 - \frac{x}{2} \right) + f_{cm}(b_i' - b)h_i \left(h_0 - \frac{h_i'}{2} \right)$$
$$= 13.5 \times 200 \times 280 \times \left(640 - \frac{280}{2} \right) + 13.5 \times (400 - 200) \times 100 \times \left(640 - \frac{100}{2} \right)$$
$$= 5.37 \times 10^8 \text{ N} \cdot \text{mm} > Ne = 5.33 \times 10^8 \text{ N} \cdot \text{mm}$$

The calculation indicates that it's appropriate to adopt $A_p = 760 \text{ mm}^2$, which meets the bearing capacity demand.

j. Check the oblique sectional bearing capacity. Verify the dimension of section

$$0.25 \ f_c b h_{01} = 0.25 \times 12.5 \times 200 \times 665$$
$$= 4.156 \times 10^5 \text{ N} > V_{\max} = 1.997 \times 10^5 \text{ N (section meets requirement)}.$$

Check the oblique sectional bearing capacity of the retrofitted beam. The stirrups of original beam are $\phi 6@200$, bent bar is $1\Phi 22$, and checked according to Eq. (3.27). Shearing resistance of the beam, V_u', is

$$V_u' = 0.07 \ f_c b h_0 + 1.5 \ f_y \frac{A_{sw}}{S} h_0 + 0.8 \ A_{sb} f_y \sin \theta$$
$$= 0.07 \times 12.5 \times 200 \times 665 + 1.5 \times 210 \times \frac{28.3 \times 2}{200} \times 665$$
$$+ 0.8 \times 310 \times 0.707 \times 380.1$$
$$= 2.42 \times 10^5 \text{ N} > V_{\max}$$

Calculation shows the section dimension and reinforcements of original beam can meet the shearing resistance requirement after the increase of the load.

k. Calculate the deflection and stretching value of the retrofitted beam. When the prestressed reinforcements are stretched, the deflection of beam, f_1, as well as the invert arch caused by prestress, f_p, is

$$E_c I_0 = 25.5 \times 10^{10} \text{ kN} \cdot \text{mm}^2 = 25.5 \times 10^4 \text{ kN} \cdot \text{m}^2$$

When the prestressed reinforcements are stretched, the residual deflection of beam can be obtained according to Eq. (3.22a), while

$$f_1 = \frac{5}{48} \alpha \frac{M_0 L^2}{B_1} = \frac{5}{48} \times 1.1 \times \frac{199.46 \times 9^2}{0.5 \times 25.5 \times 10^4} = 0.015 \text{ m}$$

The invert arch caused by prestress, f_p, is

$$f_p = -\frac{M_p L^2}{12 B_p} = -\frac{141 \times 9^2}{12 \times 0.75 \times 25.5 \times 10^4} = -0.0049 \text{ m}$$

The deflection of beam after retrofitted loads acted on it, f_2, is

$$f_2 = \frac{5}{48} \frac{M_p L^2}{B_2} = \frac{5}{48} \times \frac{(557.2 - 199.46) \times 9^2}{0.75 \times 25.5 \times 10^4} = 0.0158 \text{ m}$$

Thus ultimate deflection of the retrofitting beam is

$$f = 0.015 + 0.0158 - 0.0049 = 0.026 < \frac{L}{300} = 0.03 \text{ m (requirement is met)}.$$

l. Calculate vertical stretching quantity. In this problem horizontal bars are vertically stretched at two points. Vertical stretching quantity, ΔH, can be obtained from Eq. (3.39),

$$\Delta H = f + \sqrt{L_1 \Delta L} = f + \sqrt{L_1 \frac{\sigma_{con}}{E_s} L}$$

Because the stretching points are close to the end, approximately adopt $f = 0$, obtain ΔH from Fig. 3.27.

$$\Delta H = \sqrt{1500 \times \frac{425}{1.8 \times 10^5} \times 8500} = 174 \text{ mm}$$

Example 3.4   A roof beam is shown in Fig. 3.28, distributed dead load on original beam is 12.04 kN/m, and distributed live load is 4.36 kN/m. Later because of adding stories, load is increased by $q = 15.0$ kN/m. Try to design the retrofitting of this beam (all the loads are factored load).

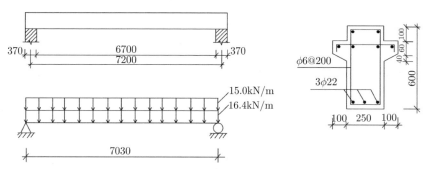

Fig. 3.28   Load and section of certain roof beam.

Solution:

a. Conditions of the original beam. Tensile main bars are $3\phi22$, $A_s = 1140$ mm^2, $f_y = 210$ N/mm^2, strength grade of concrete is C25, $f_{cm} = 13.5$ N/mm^2.

Normal-sectional bearing capacity is calculated as rectangular and without regard to the flange area.

b. Retrofitting method. Adopt lower-supported transverse frapping method (Fig. 3.29). Two ends of the prestressed reinforcements are anchored by high strength bolt bonding. Before screwing down the high strength bolts, chisel lateral surfaces of beam to be coarse in the regions touched with steel plates, and coat some epoxy mortar, then screw the high

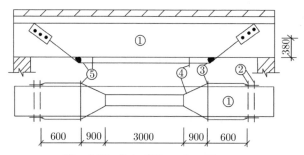

Fig. 3.29   Retrofitting of roof beam:
① original beam; ② high strength bolt; ③ support; ④ U-shaped bolts; ⑤ prestressed reinforcements.

strength bolts down. The prestressed reinforcements work minimally. Adopt steel of grade II as prestressing tendons, $f_{pyk} = 315$ N/mm^2, $f_p = 290$ N/mm^2.

c. Calculate internal forces. When the beam is retrofitted, the bending moment on the beam is

$$M_0 = \frac{1}{8}ql^2 = \frac{1}{8} \times 12.04 \times 7.03^2 = 74.38 \text{ kN} \cdot \text{m}$$

Under overall load effect

$$M_{\max} = \frac{1}{8} \times (12.04 + 4.36 + 15.0) \times 7.03^2 = 194 \text{ kN} \cdot \text{m}$$

$$V_{\max} = \frac{1}{2} \times (12.04 + 4.36 + 15.0) \times 7.03 = 110.37 \text{ kN}$$

d. Estimate the sectional area of reinforcing tendon, approximate height of compression region, x, is

$$x = \left(1 - \sqrt{1 - \frac{2M}{f_{cm}bh_{01}^2}}\right) h_0$$

$$= \left(1 - \sqrt{1 - \frac{2 \times 194 \times 10^6}{13.5 \times 250 \times 565^2}}\right) \times 565 = 113 \text{ mm}$$

Bending moment resisted by prestressied reinforcements, ΔM, is

$$\Delta M = M_{\max} - A_s f_y \left(h_0 - \frac{x}{2}\right)$$

$$= 194 \times 10^6 - 1140 \times 210 \times \left(565 - \frac{113}{2}\right) = 7.2 \times 10^7 \text{ kN} \cdot \text{m}$$

Thus,

$$A_p = \frac{\Delta M}{f_{py}\left(h + a_p - \frac{x}{2}\right)} = \frac{7.2 \times 10^7}{290 \times \left(600 + 30 - \frac{113}{2}\right)} = 435 \text{ mm}^2$$

Choose 2Φ20, $A_p = 628 \text{ mm}^2$.

e. Determine the tension control stress and calculate loss of prestress. According to Table 3.4, adopt the tension control stress as

$$\sigma_{con} = 0.85 \, f_{pyk} = 0.85 \times 315 = 268 \text{ MPa}$$

Calculate the loss of prestress. Two ends of the prestressed reinforcements are anchored by welding, so $\sigma_{l1} = 0$.

Calculate the loss of prestress caused by friction at lower supporting points, σ_{l2}. Known from Fig. 3.29, the angle between diagonal brace and longitudinal axis is 0.565 rad, assume the friction coefficient $\mu = 0.25$. Then substituting them into Eq. (3.43) gives

$$\sigma_{l2} = \sigma_{con}(1 - e^{-\mu\theta}) = 268(1 - e^{-0.25 \times 0.565}) = 35.3 \text{ MPa}$$

The loss of prestress caused by stress relaxation can be obtained by Eq. (3.44).

$$\sigma_{l4} = 0.05\sigma_{con} = 13.4 \text{ MPa}$$

Hence,

$$\sigma_l = \sigma_{l2} + \sigma_{l4} = 35.3 + 13.4 = 48.7 \text{ MPa}$$

f. Calculate prestress internal forces.

$$N_p = \sigma_{con} A_p = 268 \times 628 = 2.2 \times 10^5 \text{ N}$$

$$M_p = N_p\left(\frac{h}{2} + a_p\right) = 2.2 \times 10^6 \times \left(\frac{600}{2} + 30\right) = 7.26 \times 10^7 \text{ N} \cdot \text{mm}$$

g. Check the normal sectional strength of original beam according to eccentric compressive members. External bending moment resisted by beam is

$$M = M_{\max} - M_p = 194 - 72.6 = 121.4 \text{ kN} \cdot \text{m}$$

$$e_0 = \frac{M}{N} = \frac{121.4 \times 10^5}{2.2 \times 10^5} = 552 \text{ mm} > 0.3 \, h_0$$

Thus,

$$e_a = 0, \quad e_i = e_0 + e_a = 552 \text{ mm}$$

Adopt $\eta = 1$, $\eta e_i = 552 > 0.3\, h_0$, check according to large eccentric compression firstly,

$$x = \frac{N_p + f_y A_s}{f_{cm} b} = \frac{2.2 \times 10^5 + 1140 \times 210}{13.5 \times 250} = 136 \text{ mm} < \xi_b h_0 \quad \text{(large eccentricity)}$$

$$e = \eta e_i + \frac{h}{2} - a_s = 552 + \frac{600}{2} - 35 = 817 \text{ mm}$$

$$Ne = 2.2 \times 10^5 \times 817 = 1.8 \times 10^8 \text{ N} \cdot \text{mm}$$

$$f_{cm} bx \left(h_0 - \frac{x}{2} \right) = 13.5 \times 250 \times 136 \times \left(565 - \frac{136}{2} \right)$$
$$= 2.28 \times 10^8 \text{ N} \cdot \text{mm} > Ne = 1.8 \times 10^8 \text{ N} \cdot \text{mm}$$

The calculation shows that adopting A_p as 2Φ20 meets the bearing capacity requirement.

h. Check the oblique sectional bearing capacity. Check the shearing resistance of the retrofitted beam using Eq. (3.28).

$$V \leqslant V'_u + (\sigma_{con} - \sigma_l) A_p \sin \theta + 0.05\, N_p$$

where

$$V'_u = 0.07\, f_c b h_0 + 1.5\, f_y \frac{A_{sv}}{s} h_0$$
$$= 0.07 \times 12.5 \times 250 \times 565 + 1.5 \times 210 \times \frac{28.3 \times 2}{250} \times 565$$
$$= 1.64 \times 10^5 \text{ N} > V_{\max} = 1.1 \times 10^5 \text{ N}$$

The calculation shows the shearing resistance of the original beam can ensure the safety of the oblique section.

i. Calculate the prestressing effect and the frapping value. The calculation of prestressing effect is omitted here. For lower-supported two-point frapping, frapping value can be obtained by Eq. (3.26a),

$$\Delta L_0 = \frac{\sigma_{con}}{E_s} L_0 = \frac{268}{200000} \times 4800 = 6.43 \text{ mm}$$

To get ΔS_L, close shortened deformation ΔS should be obtained firstly. Before stretching, the beam is only affected by dead load; by using Eq. (3.30a) and Eq. (3.30b), nonuniformity coefficient and strain of the steel stress can be computed.

$$\sigma_{s1} = \frac{M_1}{0.87 h_0 A_s} = \frac{74.38 \times 10^6}{0.87 \times 565 \times 1140} = 133 \text{ N/mm}^2$$

$$\phi_1 = 1.1 - \frac{0.65 f_{fk}}{\rho_{ct} \sigma_{s1}} = 1.1 - \frac{0.65 \times 1.5}{\dfrac{1140}{0.5 \times 250 \times 600} \times 133} = 0.62$$

The horizontal project length of prestressed reinforcements below the natural axis is 6 m, thereupon,

$$\Delta S_1 = \phi_1 \frac{\sigma_{s1}}{E_s} L = 0.62 \times \frac{133}{2.1 \times 10^5} \times 6000 = 2.36 \text{ mm}$$

When the stretching is finished, the bending moment acted on original beam, M_2, is

$$M_2 = M_1 - M_p = 74.38 - 72.6 = 1.78 \text{ kN} \cdot \text{m}$$

because

$$M_2 < M_{cr} = 0.235 \, bh^2 \, f_{tk} = 0.235 \times 250 \times 600^2 \times 1.5 = 31.7 \text{ kN} \cdot \text{m}$$

Thus adopt $\Delta S_2 = 0$.

The close shortened deformation ΔS is given by Eq. (3.31)

$$\Delta S = \Delta S_1 - \Delta S_2 = 2.36 \text{ mm}$$

Because of adopting welding to anchor, the deformation of anchorage devices is 0, and then the value of ΔS_L is obtained

$$\Delta S_L = \Delta S + \frac{\sigma_{con}}{E_s} 2L_2 + 2a$$

$$= 2.36 + \frac{268}{2.0 \times 10^5} \times 2 \times 600 = 3.97 \text{ mm}$$

At last, transverse frapping value, ΔH, is obtained as

$$\Delta H = \sqrt{L_1(\Delta L_0 + \Delta S_L)} = \sqrt{900 \times (6.43 + 3.97)} = 96.7 \text{ mm}$$

From the calculation above, it is noted that the transverse frapping value considerably influences the stretching effect of prestressed reinforcements. If the influence is ignored, the tension stress will decrease brought down by 50%, which results in greatly diminishing the effect of retrofitting. This type of influence increases along with the increase of sectional area ratio of prestressed reinforcements to original steel bars, and decreases along with the decrease of loads acted on the original beam when constructing. Hence, it is important to consider the effect of closing shortened deformation in retrofitting calculation of some members.

In practical work, prestressing retrofitting is flexible and diverse, and sometimes incorporating other retrofitting methods is necessary. Hereinafter an example is given to illustrate.

Shanghai Arts and Crafts Service was built in 1915. It was two stories when it was built; later it was increased to four stories by use of coating portal frame construction with high columns (three stories high). The crossbeam of the frame adopted thin webbed girder, spanning 23 m, of which sectional height is 5 m, and thickness of web was 250 mm. At two ends of the girder, small door opening was established as the access to the fourth floor, and the third floor was a large-space market without columns. Thereafter, two more stories were added by adding columns on the beam, and part of the structure became six stories high.

Because of adding stories again and again, the depth of roof was deepened whenever water leaked; as a result of this, it was found that the depth of roof concrete reached 380 mm by coring method. Because all the load increase are delivered by the crossbeam, three beams in four cracked severely, the length of oblique cracks reached 4/5 depth of the beam, and the widest cracks reached 3 mm.

Owner requests: a) retrofit of crossbeam; b) establish a door opening of 2 m $\times$ 2.4 m at the middle of the beam; c) integrally increase to six stories; d) set up a water tank of 60 kN and a pumping room on the roof of the sixth floor.

After repeated discussion, the comprehensive retrofitting proposal containing retrofitting the crossbeam and releasing the load on beam was adopted.

The method of retrofitting the crossbeam was to cast a W-shaped truss closely combined with the original thin webbed beam at each side of the thin webbed beam. 8Φ30 curved prestressed reinforcements were established in the longitudinal direction, and the prestress was transferred to the crossbeam by W-shaped truss. Mechanical stretching was adopted. At two ends of the beam 8Φ16 vertical prestressed stirrups were established. The ends of the stirrups were anchored at the upper and lower flange of the beam and transverse frapping were used to stretch.

The method of releasing the load on the beam was to replace the partition walls on the fifth and sixth floors by light-weight materials, and replace original floor slabs with ceramic concrete floor slabs.

The performance of the building in later service period has been good based on the above comprehensive retrofitting method.

3.2.5   Sticking Steel Reinforcement Method

1. Introduction

Sticking steel reinforcement is used to retrofit structures through sticking steel plates to the exteriors of members. Conventional adhesive, the *structural adhesive*, is an epoxy-based material mixed with proportional curing agent, flexibilizer and plasticizer.

In recent years, sticking steel reinforcement method has been developed for structural retrofitting and repairing. Steel plate pasted in concrete tensile region will improve the tensile and flexural strength remarkably. Currently, this method has become so mature that the United States has established a construction standard for structural adhesive, Japan has compiled a structural adhesive quality test standard, and China has also put this approach into a *Technical Specification for Strengthening Concrete Structures*.

Sticking steel reinforcement method has won the favor of engineering technical personnel and owners because it has following advantages compared with traditional retrofitting methods:

a. Short construction period, little or no downtime due to fast hardening of adhesive, which has special appeal to owners.

b. Simple process, fast and convenient construction, without open flame, which is extraordinarily suitable for high fire-protection workshop.

c. Since the strength of adhesive is higher than concrete itself, attachment and original members can work cooperatively without stress concentration in concrete.

d. Pasted steel plates occupy small space and scarely increase the section dimension and weight of the strengthened member. Therefore, structural clearance and member shape will not be altered.

2. Properties of structural adhesive

(1) Components of structural adhesive
Structural adhesive is an epoxy-based material and its advantages are as follows:

a. Epoxy resin has high adhesiveness and good bond strength with most materials such as metal, concrete, ceramics and glass.

b. Epoxy resin has good processing property and stable storage performance. It can be prepared as thick paste or thin grouting materials, whose curing time would be adjusted appropriately according to needs.

c. Cured epoxy adhesive has excellent physical and mechanical properties, corrosion resistance capacity, and small curing contraction.

d. Epoxy material has low cost, nontoxicity and plentiful material resources.

Only by adding the curing agent can epoxy resin be cured. Because curing epoxy resin alone is brittle, flexibilizer and plasticizer are needed before curing to improve plasticity, ductility, shock strength and freeze resistance.

Epoxy curing agent types include ethylenediamine, diethanoltriamine, and triethanoltetraamine. Conventional plasticizers, which do not participate in the curing reaction, are dibutyl phthalate, dioctyl phthalate, tributyl phosphate, etc. Flexibilizers (active plasticizers) participate in curing, and generally are polyamide, butadieneacrylonitrile rubber, polysulfide rubber, etc. In addition, in order to thin the consistency of epoxy resin, a diluent, such as acetone, benzene, toluene or xylene is needed.

Currently, all structural adhesive sold on the market has two components. Component A is an epoxy resin with added plasticizer, a kind of modifier and filler. Component B is made from a curing agent and other auxiliary component. According to a certain proportion of components A and B, structural adhesive will be prepared.

(2) Physical and mechanical properties testing of structural adhesive

Whether steel plate can effectively work with original beams and fill the retrofitting role mainly depends on the shear strength and tensile strength of adhesive between steel plate and beam.

Structural adhesives JGN I and II are recommended in *Technical Specification for Strengthening Concrete Structures*. Their bonding strengths are shown in Table 3.6.

Table 3.6   Bonding strengths of structural adhesive JGNs

Bonded Material	Failure	Shear Strength (MPa)			Axial Tensile Strength (MPa)		
		Test value f_t^0	Nominal value f_{tk}	Factored value f_t	Test value f_t^0	Nominal value f_{tk}	Factored value f_t
Steel-Steel	Adhesive	$\geqslant 18$	9	3.6	$\geqslant 18$	16.5	6.6
Steel-Concrete	Concrete	$\geqslant f_v^0$	f_{cvk}	f_{cv}	$\geqslant f_{ct}^0$	f_{ctk}	f_{ct}
Concrete-Concrete	Concrete	$\geqslant f_v^0$	f_{cvk}	f_{cv}	$\geqslant f_{ct}^0$	f_{ctk}	f_{ct}

Bonding strength of structural adhesives should be tested and disqualified adhesives must not be used. Currently, there is no such test specification in the construction industry. Manufacturers in China test their products in accordance with the national *Test Method for Shear Impact Strength of Adhesive Bonds* and *Test Method for Tensile Strength of Adhesive Bonds*.

In fact, qualified adhesive strength is far higher than that of concrete. Tests show that in bond-shear test and bond tensile test, failure occurred in concrete.

(3) Stick steel reinforcement beam tests

Tongji University, Research Institute of Structural Engineering of the China Academy of Building Research, Southeast University, Shanghai Institute of Architectural Science and others have conducted various forms of stick steel reinforcement beam tests. Based on test data at home and abroad, the following conclusions apply:

a. Reinforcing with sticked steel plate can increase the cracking load of original beam. The plate, located in the margin of the beam, can effectively control the concrete tensile deformation, and is far more valid than the reinforcing bar in the original beam in improving the beam crack resistance.

b. It enhances flexural stiffness and reduces deflection.

c. It improves bearing capacity. The upgraded range shall be increased along with the increasing section area of sticked steel plate and the increasing reliability of the steel plate anchorage.

3. Failure features and force analysis of stick steel reinforcement beam

(1) Failure features

Tests show that steel plate stuck at the bottom of a beam can achieve yield strength when damaged. In proper reinforced concrete beams, as load increases, the reinforced beam is destroyed when concrete has been crushed after steel plate and rebar yield.

However, some experiments showed that when such reinforced beam was destroyed, the steel plate was still below the yield strength. Destruction is due to avulsion of the concrete and the end steel plate. Stress in rebar is suddenly increasing and entering the strengthening stage as soon as the end plate is loose, then brittle failure will occur without obvious signs.

(2) Force analysis

a. Avulsion reasons. In the second case mentioned above, why is the plate avulsed before yielding as the strength of the adhesive is rather high? Upon analysis the main reasons are the following:

a) Compared with the rebar in concrete, stick plates have more disadvantages. Tensile stress in the plates is only balanced with the single-sided bond stress.

b) The force couple, formed by the composite force of plates and the bond stress are not in a line, make steel plates deform in the opposite direction of the beam bending and avulse the plate.

c) The bonding layer is under shear and tensile combined stress.

d) Lack of anchorage between end plates and concrete.

e) Adhesive quality and construction process affect the bond quality.

b. Stress hysteresis in steel plates. Generally, structures are retrofitted without unloading. So, certain stress has existed in rebar of the original beam, whereas it may start to be generated under new adding loads in steel plate. Therefore, before steel plate yields, rebar has already yielded, and deflection and cracks of the beam will develop fast when the sticking steel plate yields.

4. Calculation of bearing capacity and specifications

When bond strength of adhesive has met the *Technical Specification for Strengthening Concrete Structures* through testing, the bond strength can be obtained from tables.

(1) Calculation of retrofitting tensile region of flexural members

a. Calculation of bearing capacity. The following expressions are proposed to compute the bearing capacity of the retrofitted beam:

$$f_{cm}bx = f_y A_s + f_{ay} A_a - f'_y A'_s \tag{3.47a}$$

$$M_u = f_{cm}bx \left(h_{01} - \frac{x}{2} \right) + f'_y A'_s \left(h_{01} - a'_s \right) \tag{3.47b}$$

where f_{ay} is the factored tensile strength of the stick steel plate, A_a is the section area of the steel plate, A_s and f_y are respectively for section area and factored tensile strength of longitudinal tension reinforcement in original beam, A'_s and f'_y are respectively for section area and factored tensile strength of longitudinal compression reinforcement in original beam, a'_s is the covering layer thickness of compression reinforcement.

b. Calculation of anchorage length. Anchorage length L_1 refers to the stick steel plate extension length outside the beam section that has no need to be retrofitted. When stress distribution coefficient equals two, the anchorage length of tensile steel plate is given as follows:

$$L_1 \geqslant \frac{2 f_{ay} t_a}{f_{cv}} \tag{3.48}$$

where t_a is the thickness of tensile sticked steel plate, f_{cv} is the factored shear strength of concrete, obtained from in Table 3.7.

Table 3.7   Concrete shear strength

Grade / Types	C15	C20	C25	C30	C35	C40	C45	C50	C55	C60
Test Value f^0_{cv}	2.25	2.70	3.15	3.55	3.90	4.30	4.65	5.00	5.30	5.60
Nominal Value f_{cvk}	1.70	2.10	2.50	2.85	3.20	3.50	3.80	3.90	4.00	4.10
Factored Value f_{cv}	1.25	1.75	1.80	2.10	2.35	2.60	2.80	2.90	2.95	3.10

Apart from meeting the above formula, anchorage length should meet constructional demands cited in the following text.

If the anchorage length is unable to meet Eq. (3.48), U-shape plates could be pasted on the end plate to strengthen anchorage, or expansion bolt could be added in. In the case of adopting U-shape plates, anchorage length should meet the following requirements:

$$\text{When} \quad f_v b_1 \leqslant 2f_{cv}L_u, f_{ay}A_a \leqslant 0.5f_{cv}b_1L_1 + 0.7nf_v b_u b_1 \tag{3.49}$$

$$\text{When} \quad f_v b_1 > 2f_{cv}L_u, f_{ay}A_a \leqslant (0.5b_1L_1 + nb_uL_u) f_{cv} \tag{3.50}$$

where n is the number of U-shape plates in each end, b_u is the width of U-shape plates, L_u is the anchorage length of each U-shape plate in the flank of beam, f_v is the factored bond-shear strength between steel plates, based on Table 3.6.

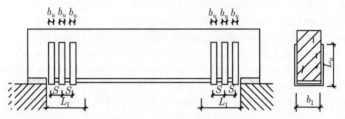

Fig. 3.30   Arrangements of U-shape stirrups in beam.

(2) Calculation of strengthening the shear capacity in oblique section

Steel ribbons could paste vertically in local parts to enhance the shear strength. Fig. 3.31(a) shows the method using parallel installed U-shape, Fig. 3.31(b) shows the method using stick steel ribbons obliquely and screw expansion bolts to anchor.

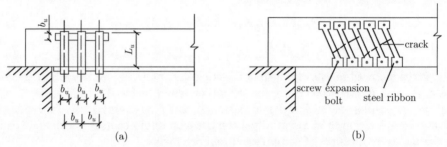

Fig. 3.31   Anchorage scheme of shear steel ribbons.

Thus, shear resistance of the oblique section shall be computed by:

$$V \leqslant V_0 + 2f_{ay}A_{al}L_u/S \tag{3.51}$$

Meanwhile, the following should be satisfied:

$$L_u/S \geqslant 1.5 \tag{3.52}$$

where V is the factored shear value of oblique section, V_0 is the factored shear value of the original beam oblique section, A_{al} is the area of each steel ribbon, S is the space length between the axial lines of two adjacent ribbons.

(3) Calculation of strengthening the compressive region of bending members

Sticking steel plates in two sides of compressive region shall apply to beams that lack compressive strength and were subjected to the section-enlarging method (Fig. 3.32).

The compressive height x is given as follows:

$$f_{y0}A_{s0} - f'_{y0}A'_{s0} - f'_{ay}A'_a = f_{cm0}b_0 x \tag{3.53}$$

The bearing capacity can be computed by:

$$M_u = f_{cm}bx\left(h_0 - \frac{x}{2}\right) + f'_y A'_s\left(h_0 - a'_s\right) + 0.9f'_{ay}A'_a\left(h_0 - \frac{b_1}{2}\right) \tag{3.54}$$

where, f'_{ay} is the factored compressive strength of sticked plate, A'_a is the section area of the plate, b_1 is the width of the plate.

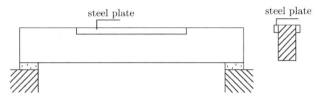

Fig. 3.32   Retrofit scheme of beam stuck with steel plates in compression region.

5. Construction provisions

Construction of members retrofitted by sticking steel plates shall comply with the following requirements:

a. The strength grade of concrete pasted with the plates should not be less than C15;

b. Range of the steel plates thickness shall be 2~4 mm. They should be stratified and pasted when the computed thickness is larger than 4 mm.

c. If retrofitted in the compressive region, the width of the steel plate may not be larger than 1/3 of the beam height.

d. 200 t (t refers to the thickness of plates) or 600 mm, whichever is greater, shall be taken as the anchorage length in a tensile region. For the compressive region, the anchorage length is the greater one of 160 t and 480 mm. Additional anchorage measures such as using bolts may be taken in earthquake resistant structures, long-span constructions and structures likely to bear cyclic reverse loads.

e. Steel plate surface should be plastered with M15 cement mortar, the thickness of which should not be less than 20 mm for beams and 15 mm for slabs.

f. To retrofit the tensile regions in supports of continuous beams exposed to negative moment, different methods could be chosen based on whether there are some obstacles such as column:

a) If not, steel plates could be pasted on top surface (Fig. 3.33).

b) For obstacles on the top surfaces but not the sides of beam, steel plates should be pasted on the sides of upper parts in beams (Fig. 3.34).

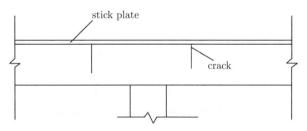

Fig. 3.33   Retrofit of continuous beam with steel plates on top surface.

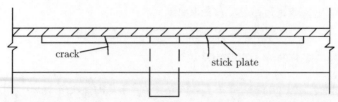

Fig. 3.34 Retrofit of continuous beam with steel plates on flank.

c) Steel plates should be stuck on the upper flank sides when both the top surface and sides of the continuous beam have stumbling blocks. At the root of beam, steel plate should be folded round the column at the slope of 1:3. The folded site should be padded with tie plates and fastened with anchor bolt (as shown in Fig. 3.35) whereas the space between the steel and beam should be filled with epoxy mortar.

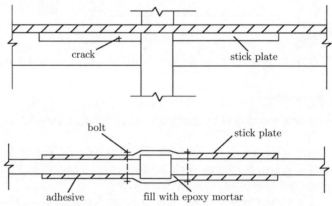

Fig. 3.35 Retrofit of frame beam with steel plates at support.

6. Construction of sticking steel plates

(1) Construction procedures
Construction procedures shown in Fig. 3.36 should be followed in the sticking steel reinforcement method.
Treatment of concrete surface, show as below:

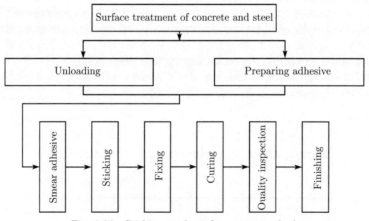

Fig. 3.36 Sticking steel reinforcement method.

a. For the bonding area of original concrete members, first brush surface grease stains with scrubbing brushes dipped in efficient detergent and flush with clear water, then polish and remove the surface layer of 2~3 mm thickness to reach the bare surface. Finally, blow away dust particles using compressed air without grease. If concrete surface is not dirty and worn the bonding area could be polished directly and the surface layer of 1~2 mm thickness removed. After flushing away dust particles with compressed air or clear water, use absorbent cotton with acetone to wipe the surface.

b. The bonding area of new concrete should be brushed with a wire brush, then brushed with a scrubbing brush dipped in efficient detergent or flushed with pressurized water. Adhesive is smeared till concrete has dried completely.

c. Concrete members that are less than 3 months old or in high humidity should be dried artificially.

(2) Treatment of steel surface before sticking

a. Removing rust and coarsening should be done. Steel plates with little or no rust could be polished with grit blast, coated abrasive or grinding wheel till appearance of metallic luster. Coarser is better. The polished grain should be perpendicular to the direction of forces applied to the steel plate. After polishing, wipe with absorbent cotton dipped in acetone. After submersion in hydrochloric acid for 20 minutes to remove rust, steel plates that rusted deeply should be flushed with limewater to counteract the acid ions, and then polished using grinding wheel.

b. If possible, the retrofitted member should be unloaded before sticking. In the case of jack unloading, multi-point lifting should be used for the beam bearing uniform loads for a main girder with a secondary beam, a jack of appropriate size based on standards (and will not produce cracks) should be installed below each secondary beam.

(3) Adhesive preparation

Structural adhesive JGN composed of two components should be prepared according to instructions for product use and tested. Remove all grease, dirt, and rainwater from the adhesive container and mix the adhesive in one direction only.

(4) Sticking steel plates

a. Adhesive should be smeared on the processed concrete and steel surface using a spatula to a thickness of 1~3 mm, thick in the center and thin at the edge. The steel plate should be attached at a preset position. For attachment at elevation, one layer of dewaxing glass fiber fabric may be added to prevent flowing. If no hollow sound occurs when the surface is knocked with a hammer, the steel plates are attached to the concrete; otherwise they should be removed and attached again.

b. Attached plates should be clamped or fixed with strutting at proper pressure and excess cement extruded from the edges of steel plates should be removed.

(5) After sticking

a. Structural adhesive cures at normal temperature. If temperature is maintained at 20°C or higher, clamping or strutting can be removed after 24 hours and the retrofitted member can bear force after 3 days. If temperature is below 15°C, artificial heat should be applied with an infrared lamp.

b. Mortar should be stuccoed on the plate surface. For plates with large surface areas, installing a layer of steel wire or pea stones will aid bonding.

(6) Quality inspection

a. Bond compaction can be checked by ultrasonic waves or lightly knocking with a hammer after removing clamps or strutting. Bonded steel plates are ineffective and should be removed and restuck if the bonding area in the anchorage area is less than 90% (70% for nonanchorage areas).

b. For major projects, a sample should be taken for load testing to confirm the effectiveness of the retrofit. The design should meet the requirements for structural deformation and cracking under normal load.

7. Example for sticking steel reinforcement

Example 3.5 Two groups of 185 m long heavy-duty crane girders whose tonnage is respectively 25 t and 50 t had to be retrofitted for serious damage in the Anshan steel factory after a long operational life. Inspection found that the crane girder with 10~60 mm carburized depth was densely covered with more than 50 cracks of 0.5~10 mm width and 150~1200 mm length. Covering layer had been peeled off in four places of the beam bottom, where the major rebars had been corroded, rusted, and even broken. There were 17 hollows 10~150 mm high and with area of 50~600 cm^2.

Solution:

a. Retrofitting method and process.

After identification, steel sticking reinforcement was chosen to retrofit the crane girder, and epoxy resin perfusion was used to patch the cracks. A_3 steel plates which were 2~3 mm thick, 800 mm wide, 6600 mm (or 3000 mm) long were attached to beams with rust-broken reinforcing bars. Four layers of epoxy glass fiber reinforced plastic that were 900 mm wide were pasted at the end of plates near support to enhance the bonding of steel and bearing capacity of oblique section.

Process: coarsening concrete—rust removing of the exposed steel using steel brush—clearing dust with compressed air—coating YJ-302 concrete finishing agent—plastering high strength and fast hardening mortar of M15—wiping the surface with acetone when $f_c \geqslant$ 10 MPa—smearing the structural adhesive on the surface of concrete and dealt steel plates—sticking plates and fixing—pasting four layers of epoxy glass fiber reinforced plastic after 16 hours curing—48 hours curing —regular service.

b. Identification of retrofitting.

Five analog experiments had been done before retrofitting, four beams of which were reinforced with 1.2 mm thick, 1.86 m long steel plates and two layers of epoxy glass fiber reinforced plastic were pasted and coiled at the end of steel. Fatigue test and static test were applied on one beam, while only static test on the others. Tests showed that bearing capacity of the former was increased 2.15 times, the latter were increased 1.24~3.7 times.

Several years after retrofitting, this project is still operating normally.

3.3   Concrete Column Retrofitting

Reinforced concrete columns are the most popular in China along with steel columns and brick columns. This chapter covers retrofitting and retrofitting design of reinforced concrete columns. Brick columns will be discussed in Chapter 4.

3.3.1   Problems in Reinforced Concrete Columns and Analysis

Generally, column destruction occurring suddenly without any signs is brittle failure. Therefore, we should understand the failure features and reasons. Analysis is needed to decide whether to retrofit the columns or not.

1. Failure features of concrete columns

Destruction of concrete columns could be divided to compression failure (including axial and small eccentric compression columns) and tensile failure (large eccentric compression columns).

(1) Failure of axial compression columns

When columns are subjected to relatively large external forces, longitudinal cracks appear in the direction of forces, then concrete in covering layer is shelled, peeled off, crushed and broken. This process has slight differences with different position of rebar. For example, when concrete coverage is thin and the space between stirrups is too large after concrete outside the bars has been shelled, peeled off and broken, the reinforcing bars will be buckled into a lantern shape. This failure occurs suddenly with small longitudinal deformation of the member.

(2) Failure of small eccentric compression columns

Small eccentric compression columns damage on the side with relatively large stress. Once longitudinal cracks appear in this side, this column is approaching failure, while rebars in the other side may be in compression, or in tension, but fail to reach yield point. If rebars are in tension, transverse cracks may generate before destruction. The increment in the compressive region is larger than that in tensile region, and the height of compression zone increases slightly. If rebars are in compression, no appearance phenomenon is obvious before the destruction. In a word, members under small eccentric compression are likely to fail without any obvious signs. Once longitudinal cracks appear in concrete, this member is very dangerous, and near to failure.

(3) Failure of large eccentric compression columns

Transverse cracks first generate on external column surface in tensile side and extend continuously along with load increasing. When stress of tensile bars reaches the tensile yield limit, transverse cracks develop quickly and extend to compressive region, and result in rapid decrease of compressive area. Finally, longitudinal cracks appear in the compressive zone, and destruction occurs when concrete disintegrates. In destroyed segment, transverse cracks develop widely while rebars in compressive side may reach compressive yield limit generally. Compressive bars may not reach the yield limit under certain circumstances such as smaller quantity of tensile bars or improperly positioned compressive bars (near natural axis). Ultimate bearing capacity of large eccentric compression columns depends on the quantity and strength of tensile rebars.

In summary, unlike large eccentric compression columns, axial and small eccentric compression columns are likely to fail without any obvious signs, which are brittle failures. Cracks on the tensile side of the column and peeling show that this reinforced concrete column is close to failure. Temporary support should be set up immediately and measures should be taken to retrofit.

It is important to identify mechanical characteristic of reinforced concrete columns. For large eccentric compression columns it is effective to retrofit tensile sides and compressive sides for small eccentric compression column.

2. Reasons for deficiency of concrete column bearing capacity

a. Ill-considered design (neglecting loads, relatively small section and miscalculation, etc). For instance, an inner frame structure has one layer underground and seven stories. Cracks appeared on the top of circular column in basement 3 months after completion. At that time the number of cracks increased to 15 after 10 days, while the crack width increased from 0.3 mm to 2~3 mm. After a half month, the stirrup in cracking places had a tension fracture, while the column had leaned to 1.68~4.75 cm and cracks had developed continuously. According to analysis, this is because the column was calculated as an axial compression column, not an eccentric compression column in design process. This column needed retrofit because the design limit bearing capacity is 1167 kN while the actual force has reached 1412 kN.

b. Poor construction quality. Such issues include substandard structural material and shoddy construction. Concrete strength will be significantly lower than design requirements when using impure sand, stone and unqualified cement. Take a five-story office building for example, which is an inner frame structure, 16.1 m long and 8.6 m wide. When the third floor was under construction, the strength of loose concrete in six columns in the first story was less than 10 MPa after detection. According to fault analysis, except for poor concrete pouring and curing, another reason is the adoption of non-factory certification cement. In a number of projects, columns burst in multiple places four or five years after completion. This is due to alkali aggregate reaction or periclase in aggregate, the volume of which will expand when producing brucite after absorbing water.

c. Low level of competence and lack of responsibility. These issues may cause the shortage of steel blanking length, overlapping length and anchorage length, wrong numbering of reinforcing bars and insufficient reinforcement. For example, a teaching building is a ten-story frame shear wall structure, 59.4 m long and 15.6 m wide. The column section size and reinforcement of the sixth story were applied to those of the fourth and fifth story. For the fourth and fifth story, the area of reinforcement of inner column decreased 4453 mm^2 (66% of design area) and the area of external column decreased 1315 mm^2 (39% of design area), which caused serious accident. Detailed retrofitting calculation of this example is shown in Example 3.8.

d. Poor management of construction site. At construction site, these circumstances often occur such as the rebars are bent and offset, or the form boards are bumped obliquely, or pouring concrete directly without adjustment. Take a factory for example, which is a five-story cast-in-place reinforced concrete frame structure. Lifting large components induced the second floor to tilt seriously. For another cast-in-place reinforced concrete frame structure, column forms were inclined when pouring concrete by reason of loose form, which made the longitudinal bars in columns inaccurate. When frame beams had been finished, the longitudinal bars in columns were uncovered. To ensure the thickness of column covering layer, construction workers bent the longitudinal bars in the shape of "$\llcorner\urcorner$" by error. Without timely retrofitting and strengthening the building is subject to structural danger.

e. Differential settlement. Differential settlement causes subsidiary stress in columns, which makes columns crack or lose bearing capacity. A factory building built on soft soil ground in Nanjing is 44 m long and 21 m wide. The superstructure is a monolayer bent structure composed of reinforced concrete columns and roof trusses, while the foundation is reinforced concrete isolated footing. Many years after completion, with the increase of output and stacking the amount of differential settlement had achieved 216~422 mm, columns tilted in different directions. Column corbel cannot bear the extra horizontal force and cracked seriously. Ultimately, the factory had to suspend production and retrofit because it cannot be used anymore.

f. Usage of commercial concrete. Generally, commercial concrete quality is better than self-mixed concrete. However, there are also some quality problems:

a) Concrete has set before arriving at construction site. Constructors re-mix this concrete with water and resume using, despite decline in concrete strength. Such circumstance occurred in an 18-story building in west Shanghai. Later, they had to retrofit these deficient columns.

b) Commercial concrete is always pumped up and lubricating mortar in conduit pipe is left in place. Sometimes, builders pour this lubricating mortar into a column, making a "mortar column". For instance, in a 26-story office building in Shanghai, such an accident occurred during building of the fifth mechanical floor, and had to be retrofitted.

Other factors can make the column bearing capacity insufficient. High-temperature fire will burn through the concrete, decreasing the strength of concrete and reinforcement. Vehicle impact will severely damage the column. Retrofitting upper structure or functional changes will increase the ability of a column to withstand loads.

Appropriate retrofitting method should be chosen and then applied in time according to column appearance, checking calculation and site condition after understanding failure feature.

There are many column retrofitting methods including section-enlarging method, replacing method, sticking steel reinforcement method, prestressing method, etc. Unloading and adding struts are sometimes adopted. These methods will be illustrated below.

3.3.2 Section-enlarging Method

1. Introduction

Section-enlarging, also known as outsourcing concrete retrofitting, is a common method to retrofit columns. For enlarging section area and reinforcement of original columns, this method enhances column bearing capacity, and also reduces column slenderness ratio and improves column stiffness. Particularly in the seismic fortified areas, it could change the original strong beam-weak column structure into the strong column-weak beam structure, to enhance seismic resistance.

Specific methods include surrounded outsourcing, two-side thickening and single-side thickening, etc.

Surrounded outsourcing means adding a reinforced concrete shell around the original column (Fig. 3.37(a), (b), (c)). In Fig. 3.37(b), after the covering layer in the original column corner has been tapped out, baring the longitudinal bars, reinforcement is assembled in exterior and concrete is poured into octagon shape to improve appearance. Using this method to enhance bearing capacity of axial and small eccentric compression columns is particularly effective.

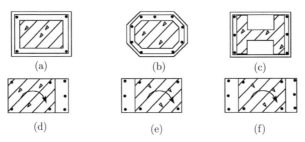

Fig. 3.37   Section enlarging methods.

When columns withstand large moment, two-side thickening in the side perpendicular to the moment plane is always adopted. Compressive side should be retrofitted (Fig. 3.37(d)) because it is unsubstantial, and so should the tensile side (Fig. 3.37(e)). It may need to be retrofitted on both sides (Fig. 3.37(f)).

A more frequently used method is shotcreting. It is simple, convenient, requiring few or no forms except for irregular columns. Shotcrete with high bond strength ($> 1.0 \text{ N/mm}^2$) could meet the repairing and retrofitting quality requirements of general structures. Repeated spraying (50 mm each time) could be used when the concrete is thick.

2. Constitution and construction requirement

In the column retrofitting design and construction process, it is important to strengthen the connection between the old and new columns, ensuring that internal forces redistribute well and they work together as new columns. Therefore, in the process of constitution design and construction, attention should be given to the following points:

a. When using surrounded outsourcing, the original surface should be coarsened and washed cleanly. Stirrups should be closed, spacing of which should follow *Code for Design of Concrete Structure* and *Code for Seismic Design of Buildings*.

b. When using two-side thickening or single-side thickening, the original surface should be coarsened, ruggedness should be larger than 6 mm, and one of the following constitution measures should be adopted:

a) When thickness of new poured concrete is rather small, retrofitted rebars are welded on original column bars using short bars (Fig. 3.38(c)). Diameter of short bars should not be less than 20 mm, the length not less than 5 d (d is the relative small diameter of new adding longitudinal bars and original longitudinal bars) and spacing between them not larger than 500 mm.

b) When thickness of new poured concrete is rather large, U-shape stirrups are used to fix longitudinal rebars by welding (Fig. 3.38(d)) or anchorage (Fig. 3.38(e)). In the case of welding, length of inconel weld is 10 d and length of double welded joint is 5 d (d is the diameter of the U-shape stirrup). Procedures of anchorage method is that drilling holes in the place of original column firstly, the distance between which and the column edge should not be less than 3 d and not less than 40 mm. The hole depth should be not less than 10 d, and the diameter should be 4 mm larger than stirrup diameter. Then stirrups are anchored in the drilled holes using epoxy grout or epoxy mortar. In addition, rivets not less than 10 mm indiameter could be anchored in holes, then weld U-shape stirrups on rivets.

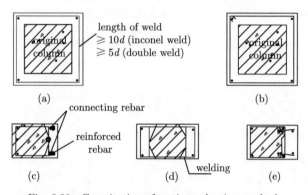

Fig. 3.38   Constitution of section enlarging method.

c. The smallest thickness of new adding concrete should not be less than 60 mm, or 50 mm when using shotcrete.

d. It is better to use transformed rebar with a diameter range from 14 mm to 25 mm.

e. New longitudinal rebar should be anchored into foundation and the top end. Tensile bars should not be cut off in floor slab, while 50% compressive bars should transverse slab and should be compacted between the top of new poured concrete and the girder bottom.

3. Force characteristics

When retrofitting concrete columns under load, stress and strain of the new concrete lag behind the original because compression deformation exists in original columns and shrinkage and creep have occurred. Therefore, the new and the old cannot simultaneously achieve peak stress, reducing the effect of the new concrete. Along with the actual stress in columns before retrofitting, reduction degree changes, and the higher the stress, the greater the degree of reduction.

Effect of new concrete is also correlated with the ratio between post-imposed loads and residual loads. New concrete doesn't bear original loads, if the loads do not increase after retrofitting. Only when loads increase, will the stress of the new concrete increase. Therefore, the new concrete is at lower stress levels and cannot fully play its role to retrofit if original columns have relative high stress and large deformation.

Experiments show that if the combination of the old and new concrete is reliable, the old and new concrete strain increments are basically the same; deformations of the entire section are compliant with the plane cross-section assumption.

If the new concrete locates in the edges of large eccentric compression columns, strain develops faster than in the original column, which covers the new column's shortage "strain hysteresis". Furthermore, due to confinement and stress redistribution, reduction of the new concrete capacity is not significant and less than axial compression columns.

Significant stress redistribution exists between the old and new concrete in axial compression columns. Tests show that the new concrete with relatively low stress will restrain the old concrete with relatively high stress; with the increase of the stress difference the restraint is more obvious. When the concrete strain in original columns achieves 0.02, concrete does not break immediately. But this restraint cannot counteract the reduction due to strain hysteresis. These tests also indicate that, when the initial compression is 0.41~0.71 times column bearing capacity, test capacity decreases 0.18~0.21 times of calculation (calculated in accordance with respective material strength). Hence, *Specification for Retrofitting of Concrete Structures* indicates that, when the range of axial compression ratio is from 0.1 to 0.9 and concrete strain reaches the ultimate compressive strain, the stress of new concrete is far less than the factored strength f_c and ratio α between them varies from 0.99 to 0.53.

Reduction coefficient should not limit the axial compression ratio in the seismic code and unloading. To simplify calculation, α equals to 0.8 under axial compression and 0.9 under eccentric compression.

Loads acting on columns should be controlled within 60% of the ultimate bearing capacity during construction. In this condition, bearing capacity could be calculated in accordance with following methods. If it cannot meet the above requirements, unloading or prestress top bracing method could be used to reduce column stress.

4. Calculation of the section bearing capacity

Using section-enlarging method, bearing capacity should be calculated in accordance with *Code for Design of Concrete Structure*, considering the combined action of the new concrete and original column.

(1) Axial compression columns

The normal section bearing capacity shall be computed by:

$$N \leqslant \phi \left[f_c A_c + f_y' A_s' + \alpha \left(f_{c1} A_{c1} + f_{y1}' A_{s1}' \right) \right] \tag{3.55}$$

Where, N is the factored axial force acted on the retrofitted columns; ϕ is the stability factor of the retrofitted section according to *Code for Design of Concrete Structure*; A_s' and f_y' are respectively for section area and factored compressive strength of longitudinal reinforcement in original columns; A_c and f_c are respectively for section area and factored axial compressive strength of concrete in original columns; A_{c1} and f_{c1} are respectively for section area and factored axial compressive strength of the new adding concrete; A_{s1}' and f_{y1}' are respectively for section area and factored compressive strength of the new longitudinal reinforcement; α is the strength reduction coefficient of the new concrete and longitudinal reinforcement because of combined action, $\alpha = 0.8$.

(2) Large eccentric compression columns

In eccentric compression columns, if the new adding concrete and the original concrete can work together, their strength can be computed as an entire section. Fig. 3.39 shows the calculation of two-side thickening columns in ultimate limit state. According to retrofitting codes, factored strength of the new concrete in both tensile and compressive regions and the longitudinal reinforcement should be multiplied by the reduction coefficient that equals 0.9. The normal section bearing capacity can be computed by:

$$N \leqslant f_{cm} b \left(x - h_1 \right) + f_y' A_s' - f_y A_s + 0.9 \left(f_{cm1} b_1 h_1 + f_{y1}' A_{s1}' - f_{y1} A_{s1} \right) \tag{3.56}$$

$$Ne \leqslant f_{cm}b\,(x-h_1)\left(h_{01}-\frac{x-h_1}{2}\right)+f'_y A'_s\,(h_{01}-h_1-a'_s)$$

$$+\,0.9\left(f_{cm1}b_1 h_1\left(h_{01}-\frac{h_1}{2}\right)+f'_{y1}A'_{s1}\,(h_{01}-a'_{s1})\right) \qquad (3.57)$$

where, f_{cm} and f_{cm1} are respectively the factored flexural compressive strength of the new adding concrete and the original, which respectively equal $1.1f_c$ and $1.1f_{c1}$; x is the compressive height; h_1 is the thickness of adding concrete in compressive region; b and b_1 are respectively the section width of the original columns and the retrofitted ones; A_s and A_{s1} are respectively the section area of tensile reinforcement in original and retrofitted columns; f_y and f_{y1} are respectively the factored tensile strength of the tensile reinforcement in original and retrofitted columns; e is the distance between action points of axial force and reinforcement resultant force, which is:

$$e = \eta e_i + \frac{h}{2} - a$$

h_{01} is the distance from compressive edge to action point of tensile reinforcement resultant force including the original and the new addition. When the two action points are close, h_{01} equals to the effective height h_0 of original columns. a'_{s1} is the distance from the compressive edge to the action point of the new adding tensile reinforcement resultant force. a is the distance from the tensile edge to the action point of tensile reinforcement resultant force including the original and the new. Meanings of the other symbols are shown in *Specification for Retrofitting of Concrete Structures* and Eq. (3.55).

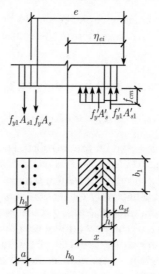

Fig. 3.39   Calculation of column limits.

Calculation of the bearing capacity of single thickening columns could make reference to Eq. (3.56) and Eq. (3.57).

By using surrounded outsourcing, calculation of the bearing capacity is complicated. To simplify calculation, factored compressive strength of the new concrete equals the flexural compressive strength f_{cm} of the original concrete, which means replacing the $0.9\,f_{cm1}$ in Eq. (3.56) and Eq. (3.57) with f_{cm}. This is based on the fact that the strength grade of the adding concrete is one level higher than that of the original concrete. The simplification indicates that $0.9\,f_{cm1}$ is a little bigger than f_{cm}.

In summary, by using surrounded outsourcing method, calculation of the right section bearing capacity is given by:

$$N \leqslant f_{cm}b_1x + f'_yA'_s - f_yA_s + 0.9f'_{y1}A'_{s1} - 0.9f_{y1}A_{s1} \tag{3.58}$$

$$Ne \leqslant f_{cm}b_1x\left(h_{01} - \frac{x}{2}\right) + f'_yA'_s(h_{01} - h_1 - a'_s) + 0.9f'_{y1}A'_{s1}(h_{01} - a'_{s1}) \tag{3.59}$$

When reinforcement of grade I or II is symmetry, Eq. (3.56) can be further simplified as follows:

$$N \leqslant f_{cm}b_1(x - h_1) + 0.9f_{cm1}b_1h_1 \tag{3.60}$$

(3) Small eccentric compression columns

In small eccentric compression columns, if the tensile bars away from axial force cannot reach the yield strength, the bearing capacity of two sides thickening columns could be computed by:

$$N \leqslant f_{cm}b(x - h_1) + f'_yA'_s - f_yA_s + 0.9\left(f_{cm1}b_1h_1 + f'_{y1}A'_{s1} - f_{y1}A_{s1}\right) \tag{3.61}$$

$$Ne \leqslant f_{cm}b(x - h_1)\left(h_{01} - \frac{x - h_1}{2}\right) + f'_yA'_s(h_{01} - h_1 - a'_s)$$

$$+ 0.9\left(f_{cm1}b_1h_1\left(h_{01} - \frac{h_1}{2}\right) + f'_{y1}A'_{s1}(h_{01} - a'_{s1})\right) \tag{3.62}$$

where, σ_s and σ_{s1} are respectively the stress in original reinforcement and retrofitting reinforcement. They can be calculated approximately equivalent:

$$\sigma_s = \frac{\xi - 0.8}{\xi_b - 0.8}f_y \tag{3.63}$$

where, ξ is the coefficient of compressive height, $\xi = \dfrac{x}{h_{01}}$; ξ_b is the coefficient of compressive height in critical failure, which equals to 0.61 for grade I steel, 0.55 for grade II steel.

For single-side thickening columns, surrounded outsourcing columns and symmetry reinforcement columns, bearing capacity can be computed as for large eccentric columns.

5. Examples

Example 3.6 A nine-story office building is a two-span reinforced concrete frame-shear wall structure. Two stories need to be added onto it. Some columns need to be retrofitted after checking. Calculations are given:

a. Original column conditions. The section size of this central column is 400 mm × 500 mm. The concrete strength grade is C20. Longitudinal reinforcement is 8 Φ 18. The story height $H = 6.5$ m. After retrofitting, this column withstands axial compression, the magnitude of which is 3600 kN.

b. Retrofitting method. It is better to use surrounded outsourcing to retrofit axial compression columns. First, coarsen concrete on four surfaces of the original column, then collocate longitudinal reinforcement and $\phi 8@200$ stirrup, and spray 50 mm C25 fine concrete.

c. Calculation of the new longitudinal reinforcement and length of the column:

$$l_0 = 1.0H = 6.5\text{m}$$

$$\frac{l_0}{b} = \frac{6.5}{0.4 + 0.1} = 13$$

According to Eq. (3.55), the area of the new adding reinforcement can be computed by ($\varphi = 0.935$):

$$A'_{s1} = \left(\frac{N}{\varphi} - f_c A_c - f'_y A'_s - 0.8 f_{c1} A_{c1} \right) \Big/ 0.8 f'_{y1}$$

$$= \left[\frac{3.6 \times 10^6}{0.935} - 10 \times 400 \times 550 - 310 \times 2036 - 0.8 \times 12.5 \right.$$

$$\left. \times (500 \times 650 - 400 \times 550) \right] \Big/ 0.8 \times 310$$

$$= \left(1.02 \times 10^6 - 1.05 \times 10^6 \right) / 248 < 0$$

The above-mentioned calculation explains that only using the shotcrete around the column can satisfy the request for retrofitting. So, only 4 Φ 14 longitudinal constitutional rebars are needed.

Example 3.7 In a five-story factory that is a frame structure, lifting large components damaged the frame form, which induced the second floor to tilt seriously. Some columns need to be retrofitted. Calculations of retrofitting a side column are given:

a. Design data. The section size of this central column is 400 mm × 600 mm. The concrete strength grade is C20. The story height $H = 5.0$ m. The original design external forces are $N_0 = 600$ kN, $M_0 = 360$ kN·m. The reinforcement is 4 Φ14 ($A_s = A'_s = 1256$ mm²). The extra design moment due to tilt is $\Delta M = 50$ kN·m.

b. Retrofitting method. Single-side thickening is applied because the moment is unidirectional. First, coarsen tensile concrete surface of original column and expose 80 mm stirrups. Then weld U-shape stirrups on original stirrups and weld the longitudinal rebar to original stirrups by short bars. Finally, spray 50 mm C25 fine concrete.

c. Computing h_{01}, η and e. From Fig. 3.40, the effective height of the retrofitted section can be computed by:

$$h_{01} = 600 - 10 = 590 \text{ mm}$$

$$\frac{l_0}{h} = \frac{1.0 \times 5}{0.65} = 7.69 < 8 \quad (\eta = 1)$$

$$e_0 = \frac{M}{N} = \frac{360 \times 10^3}{600} = 600 \text{ mm} > 0.3 \, h_{01} \quad \text{(large eccentric)}$$

$$e_a = \frac{50 \times 10^3}{600} = 83.3 \text{ mm} \quad (e_i = e_0 + e_a = 683.3 \text{ mm})$$

So,

$$e = \eta e_i + h_{01} - \frac{h}{2} = 683.3 + 590 - 325 = 948.3 \text{ mm}$$

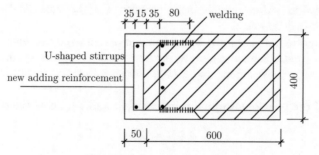

Fig. 3.40 Retrofit of column.

d. According to Eq. (3.57), the area A_{s1} of the new adding reinforcement can be computed by:

$$a_s = \frac{Ne - f'_y A'_s (h_{01} - a'_s)}{f_{cm} b h_{01}^2}$$

$$= \frac{600 \times 10^3 \times 948.3 - 310 \times 1256 \times (590 - 35)}{11 \times 400 \times 590^2} = 0.230$$

From the table, $\xi = 0.265$, $x = \xi h_0 = 0.265 \times 590 = 156.4$ mm.
  According to the Eq. (3.56):

$$A = \frac{fbx - N}{0.9 f_{y1}} = \frac{11 \times 400 \times 156.4 - 600 \times 10^3}{0.9 \times 310} = 316 \text{ mm}^2$$

$$\text{Select } 2\Phi18(A_{s1} = 509 \text{ mm}^2)$$

Example 3.8   A teaching building is ten-story frame shear wall structure. The section size and reinforcement of columns in the sixth story are to be applied to the fourth and fifth stories. The reinforcement decrease results in insufficient bearing capacity. Calculations are given as follows:

a. The story height $H = 4.0$ m. The original design axial forces $N = 4320$ kN, $M = 270$ kN·m. The section size $b \times h = 400$ mm $\times$ 450 mm. The area of needed reinforcement is 5000 mm^2, actually used reinforcement is 8 Φ 20 ($A_s = 2513$ mm^2).

b. Retrofitting method. Surrounded outsourcing is applied to retrofit columns. First, coarsen concrete on four surfaces of the original column, then collocate longitudinal reinforcement and $\phi8@200$ stirrup and spray 50 mm C25 fine concrete.

c. Computing e and η.

$$\frac{l_0}{h} = \frac{1.0 \times 4}{0.55} = 7.3 < 8 \quad (\eta = 1)$$

$$e_0 = \frac{M}{N} = \frac{2.7 \times 10^5}{4320} = 62.5 \text{ mm} < 0.3 \, h_{01} = 162 \text{ mm} \quad (\text{small eccentric})$$

So,

$$e_a = 0.12 \, (0.3 \, h_{01} - e_0) = 0.12 \times (0.3 \times 540 - 62.5) = 11.9 \text{ mm}$$

$$e_i = e_0 + e_a = 62.5 + 11.9 = 74.4 \text{ mm}$$

$$e = \eta e_i + \frac{h}{2} - a_s$$

$$= 74.4 + \frac{550}{2} - 35 = 314.4 \text{ mm}$$

From the Eq. (3.61) and Eq. (3.62), area of new adding reinforcement is computed:

$$A_{s1} = A'_{s1} = 1446.9 \text{ mm}^2$$

Select 4 Φ 22 ($A_{s1} = 1250$ mm^2)

3.3.3   Encased Steel Method

1. Introduction

Encased steel refers to retrofit the concrete column by wrapping four corners or two sides of the column with profiled steel. Its advantage is that concrete column capacity can be greatly increased along with a little increase of section size. For square or rectangle columns, rolled angles are generally wrapped on four corners and linked with horizontal batten plates to form a whole body. For circular members such as circular column and chimney, flat-rolled steel hoops with skeins are often used (Fig. 3.41).

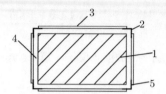

Fig. 3.41   Encased steel methods

1. original column; 2. rolled angle;
3. batten plate; 4. concrete or mortar;
5. agglomerate.

Retrofitting by bonding steel and column together by filling latex cement or epoxy mortar or fine concrete in space between them is called wet-enclosing steel method.

This method improves capacity of concrete columns and also improves ductility because of the restraints of profiled steel and batten plates.

2. Retrofitting design using wet-enclosing steel method

The lateral deformation of a column is restricted by profiled steel bonded on the column while profiled steel bears compression and moment because of lateral compression induced by concrete lateral deformation, which may result in decrease of compression strength of steel. Moreover, as with section-enlarging mentioned above, there is also stress hysteresis in profiled steel, which prevents the profiled steel to play its role completely. So, the factored strength of profiled steel should be reduced.

The normal section bearing capacity of wet-enclosing steel reinforcement column could be computed as a whole section, while the factored strength of profiled steel should be reduced. According to *Specification for Retrofitting of Concrete Structures*, the reduction coefficient of a compression rolled angle is 0.9.

Because of the small thickness of new poured concrete (or mortar) and the existence of rolled angle and batten plate, the bonding of column and new concrete is weakened. Under ultimate state, the new poured concrete is possibly flaked off and its contribution is neglected in the process of retrofitting design.

Calculation methods of section stiffness and bearing capacity of wet-enclosing steel reinforcement column are introduced below:

(1) Section stiffness

Section stiffness EI of wet-enclosing steel reinforcement column can be approximately computed by:

$$EI = E_c I_c + E_a A_a a^2 \tag{3.64}$$

where E_c and I_c are respectively for the Young's modulus of concrete in original column and the inertia moment of the original column; E_a is the Young's modulus of profiled steel; A_a is section area of profiled steel in each column side; a is the distance between centroids of area of the tensile and compressive profile steel section.

(2) Bearing capacity of axial compression columns

Bearing capacity of wet-enclosing steel reinforcement column is given as follows:

$$N \leqslant \phi \left(f_c A_c + f_y' A_s' + 0.9 f_a' A_a' \right) \tag{3.65}$$

where, f'_a is the factored compressive strength of profiled steel; A'_a is the section area of profiled steel; the remaining symbols have the same explanation as Eq. (3.65).

(3) Bearing capacity of large eccentric compression columns

Because of stress hysteresis in tensile profiled steel, its factored strength should be reduced. To simplify calculation, it is suggested that strength reduction coefficients of tensile and compressive steel are the same, equal to 0.9. Generally, profiled steel is placed symmetrically. From Fig. 3.42, calculation of normal section bearing capacity of wet-enclosing steel reinforcement column is given:

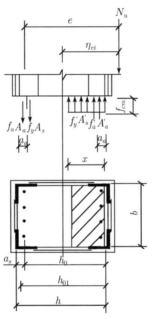

Fig. 3.42   Wet-enclosing steel reinforcement.

$$N \leqslant f_{cm}bx + f'_y A'_s - f_y A_s \tag{3.66}$$

$$Ne \leqslant f_{cm}bx \left(h_{01} - \frac{x}{2}\right) + f'_y A'_s (h_{01} - a'_s) + 0.9 f'_a A'_a (h - a_{a0} - a'_a) \tag{3.67}$$

where b is the section width of the original column; h is the section height of the original column; x is compressive height of the concrete in original column; A'_a and f'_a are respectively the section area and factored compressive strength of the compressive profiled steel; a'_a is the distance from section centroid of compressive profiled steel to the compressive edge of original column; a_{a0} is the distance from action point of the resultant force of the tensile profiled steel and reinforcement to the tensile edge of original column; the remaining symbols have the same explanation as above.

When the reinforcement in original column is symmetrical, Eq. (3.66) can be further simplified to:

$$N \leqslant f_{cm}bx$$

(4) Bearing capacity of small eccentric compression columns

Bearing capacity can be computed by:

$$N \leqslant f_{cm}bx + f'_y A'_s - \sigma_s A_s + 0.9 \left(f'_a A'_a - \sigma_a A_a\right) \tag{3.68}$$

$$Ne \leqslant f_{cm}bx \left(h_{01} - \frac{x}{2}\right) + f'_y A'_s (h_{01} - a'_s) + 0.9 f'_a A'_a (h - a_{a0} - a'_a) \tag{3.69}$$

where, σ_s and σ_a are respectively stress of tensile or small compression bars and profiled steel, which can take the same value approximately and be computed as Eq. (3.63).

3. Constitution requirement

Using encased steel method, the following requirements should be complied with:

a. The side length of rolled angle could be not less than 75 mm; the section size of batten plate could be not less than 25 mm × 3 mm, space length of which could be not larger than 20 r (r is the least radius of gyration of a single rolled angle section 2) and 500 mm.

b. Rolled angle should be continuous and not cut off in slabs. The end of rolled angle should extend to the top of foundation and be anchored with epoxy mortar or epoxy plaster. Post cap may be equipped to weld with rolled angle, ensuring that the top of rolled angle has enough anchorage length.

c. Batten plate should be close to concrete surface and connect with rolled angle through flat position welding when using epoxy to grout. After it is welded, profiled steel should be sealed by epoxy and air holes are prepared for final grouting.

d. When using latex cement in which latex should be not less than 5%, batten plate may be welded outside rolled angle.

e. 1:3 sand-cement grouts with width of 25 mm or antiseptics could be used as cover of the profiled steel.

4. Example

Example 3.9 Retrofitting Example 3.7 using wet-enclosing steel method.

a. Processing.

a) Sand the original column surface with hand-hold electric grinder until column surface becomes smooth and four corners are round angles, coarsen it using wire brush, blow clear with compressed air.

b) Smear a thin layer of epoxy resin, then stick the profiled steel which has been descaled and wiped with dimethyl benzene on surface, and clamp with a chucking appliance.

c) Stick batten plate on surface and weld.

d) Seal the profiled steel with epoxy plaster and leave air holes, then paste grout nicks in proper position, the spacing of which is 2~3 mm.

e) Check whether the grout nick leaks, and press epoxy resin into it under pressure of 0.2~0.4 MPa.

f) Spray 1:2 sand-cement grouts on the surface of the column.

b. Materials. According to constructional requirement, $4L75 \times 5$ is rolled angle, $A_s = A'_s = 741.2 \times 2 = 1482$ mm^2, $a_s = 20.3$ mm, $f'_s = 215$ N/mm^2; batten plate select 25×3, spacing is 20 $r = 20 \times 15 = 300$ mm.

c. Calculation of bearing capacity. As in Example 3.7, $M = 410$ kN·m, $N = 600$ kN, $b \times h = 400$ mm $\times 600$ mm, $A_s = A'_s = 1256$ mm^2, $\eta = 1.0$

$$e_0 = e_i = 683.3 \text{ mm} > 0.3 \, h_{01} \quad \text{(large eccentricity)}$$

$$e = \eta e_i + \frac{h}{2} - a_{s0} = 683.3 + 300 - 30 = 953.3 \text{ mm}$$

According to Eq. (3.66),

$$x = \frac{N}{f_{cm}b} = \frac{600 \times 10^3}{11 \times 400} = 136.4 \text{ mm}$$

According to Eq. (3.67), section resistance moment could be computed by:

$$f_{cm}bx\left(h_{01}-\frac{x}{2}\right)+f'_yA'_s\left(h_{01}-a'_s\right)+0.9\,f'_yA'_s\left(h-a_{s0}-a'_s\right)$$

$$=11\times400\times136.4\times\left(565-\frac{136.4}{2}\right)+310\times1256\times530$$

$$+\,0.9\times215\times1428\times(600-30-20.3)$$

$$=6.67\times10^8\ \mathrm{N\cdot mm}>Ne=5.67\times10^8\ \mathrm{N\cdot mm}$$

3.3.4   Replacing Method

1. Introduction

Concrete columns need to be retrofitted because of deficiency of bearing capacity due to fire or construction error.

2. Calculation

a. Bearing capacity of axial compressive members using partial replacing can be computed by:

$$N\leqslant\phi\left(f_cA_0+\alpha_0f_{cj}A_j+f_yA'_s\right)\tag{3.70}$$

Using whole section replacing method, it can be computed by:

$$N\leqslant\phi\left(\alpha_0f_{cj}A_j+f'_yA'_s\right)\tag{3.71}$$

where N is factored axial force after retrofitting; ϕ is stability coefficient according to *Code for Design of Concrete Structure*; f_c is the factored strength of remaining concrete; f_{cj} is the factored strength of new concrete; A_0 is the section area of remaining concrete; A_j is the section area of new concrete; α_0 is the utilization coefficient of new concrete, $\alpha_0=0.8$ when without construction supports; $\alpha_0=1.0$ when using construction supports.

b. Bearing capacity of eccentric compression concrete members using replacing method to retrofit can be calculated as the following two situations:

a) When the replacement depth of compressive concrete $h_n>x_n$, its bearing capacity can be computed according to *Code for Design of Concrete Structure* using the strength of new concrete.

b) When the replacement depth of compressive concrete $h_n\leqslant x_n$, its bearing capacity can be computed by:

$$N\leqslant f_{cm}bh_n+f_{cm0}b\left(x_n-h_n\right)+f'_yA'_m-\sigma_mA_m\tag{3.72}$$

$$Ne\leqslant f_{cm}bh_nh_{0n}+f_{cm0}b\left(x_n-h_n\right)h_{00}+f'_yA'_m\left(h_0-a'_m\right)\tag{3.73}$$

where N is factored axial force after retrofitting; f_{cm} is factored flexure compressive strength of new concrete, equal to $1.1\,f_c$; f_{cm0} is factored flexure compressive strength of concrete in original members, equal to 1.1 times the factored axial compressive strength; x_n is the height of concrete compressive region after replacement; h_n is the concrete replacement depth; h_0 is the distance from resultant force center of tensile reinforcement to the edge of compressive region; h_{0n} is the distance from resultant force center of tensile reinforcement to centroid of replacement concrete; h_{00} is the distance from resultant force center of tensile reinforcement to centroid of original concrete; A_m and A'_m are respectively the section area of longitudinal reinforcement in tensile region and compressive region; b is the width of rectangle section; a'_m is the distance from resultant force center of compressive reinforcement to the edge of compressive region; f'_y is the factored compressive strength of longitudinal compressive reinforcement; σ_m is the stress of longitudinal tensile reinforcement.

c. Bearing capacity of reinforcement concrete bending members using replacing to retrofit can be computed as the following two situations:

a) When the replacement depth of compressive concrete $h_n > x_n$, its bearing capacity can be computed using the strength of new concrete.

b) When the replacement depth of compressive concrete $h_n \leqslant x_n$, its bearing capacity can be computed by:

$$M \leqslant f_{cm}bh_n h_{0n} + f_{cm0}b\left(x_n - h_n\right)h_{00} + f_y A_m \left(h_0 - a'_s\right) \tag{3.74}$$

where M is the factored moment of the member after retrofitting; the remaining symbols have the same explanation as the former.

3. Constitutional requirements

a. The strength grade of replacing concrete is determined by calculation, and should be one level higher than original strength grade at least and not less than C25. The replacement depth is also determined by calculation, and should be not less than 40 mm for slabs, 60 mm for beams and columns, and 50 mm using sprayed concrete.

b. When using hard pebble or macadam to prepare concrete, the grain diameter of stones may not larger than 20 mm when replacing small areas, and it increases when the replacement depth is large. The maximum grain diameter may be not larger than one-third of the replacement depth and not larger than 40 mm.

4. Construction demands

a. Using this method, total or partial unloading before retrofitting is necessary. To ensure the safety of construction, structural strength in construction phase should be checked.

b. Replacement concrete construction shall comply with the following requirements:

Clear drawback in original concrete to dense part or to defined depth. Coarsen concrete surface or make cannelure, the depth of which may be not less than 6 mm and the spacing may be not larger than one-half of stirrup spacing. At the same time, scum, dust and loose stone should be removed.

Wash bond surface of the new and the old concrete clearly, one layer of high strength grade pure sanded cement grout or other interface agent should be smeared on the bond surface before pouring concrete.

c. Sprayed concrete or steel fiber reinforced concrete is preferred when using replacing concrete method. In special case that the replacement depth is rather small and it needs a moulding board, pouring and curing concrete should comply with *Code for Acceptance of Constructional Quality of Concrete Structures*.

3.4   Retrofitting of Concrete Roof Trusses

Roof trusses are vital structural members in industrial and civilian construction. The detailing of roof trusses is so complicated and the quantity so large that they are more likely to need strengthening for such cases as underestimated bearing capacity, insufficient stiffness for practical displacement, incapability for usage, over-wide cracks or serious erosion in steel bars, poor durability and even unsafe structure. The common issues and practical cases concerning about the reinforced concrete roof truss retrofit are discussed as follows.

3.4.1   Analysis of Concrete Roof Trusses

During the 1950s and early 1960s, most buildings were designed as non-prestressed concrete roof truss systems. Over-wide cracks and erosion in reinforcing bars appear in those roof trusses due to inappropriate design, construction defects and increasing service load. In addition, various problems emerge in different shapes of roof trusses (such as triangle-shaped, vaulted, trapeziform, etc.). Despite that those structures have been built in accordance with standard design procedures considering practical tests and experience, accidents occur from time to time in aged roof trusses.

1. Common cases for RC roof truss damage

(1) Non-prestressed concrete roof truss systems

The problems of design and causes of accident in non-prestressed concrete roof truss systems can be classified:

a. Among long-span roof truss structures, if the straightness of cauls is not strictly controlled or flexion and deflection occur in steel bars in the process of construction, damage initiates when the relatively longer tensile reinforcing bars become straight after loading, which is followed by cracking in lower chords and substantive deflection of global roof, and even longitudinal cracks in the direction of main reinforcement. Consequently, the defective materials erode the reinforcing bars through the cracks and covering concrete is spalled.

b. If the lower chords are welded incorrectly the problem that force points on both sides of the welding areas on reinforcing bars are not located along the same line can result in eccentricity of loading in roof trusses, which is accordingly responsible for crack occurrences. In addition, those cracked chords might break down from severe stress concentration. The collapse of the 12-meter span, 6-bay roof truss of a factory in Bachu county of Xinjiang province is a case in point. Since the lower chords were welded through binding bars, they will eventually cause large stress-concentration in roof truss and destruction of the entire structure.

c. In some cases, poor quality in construction is responsible for earlier crack emergence in concrete. For instance, the tensile and compressive strength of concrete cannot meet the demand in design, or sometimes satisfactory bars are mistaken for deformed ones.

d. Due to relatively large weight and lateral flexible stiffness, a roof truss system is prone to bear additional torsional force when lifted and straightened. Additionally, the loading direction of upper and lower chords in roof truss probably changes from the state of lifting to the state of practical use. Thus, cracks derived form hoisting will not only decrease the stiffness of roof trusses, but affect the internal force distribution under load effects as well.

e. Cracks may arise from the unsuitable arrangement of reinforcement in joints. The joints between lower chords are likely to crack before reaching design load capacity of roof truss. And this situation is usually caused by insufficient imbedding length of tensile web members into their connecting lower chords.

f. Reinforced concrete roof truss belongs to slender member structure and is made at flat cast position. Despite negligible cracks concerning probable cut-off for structural capacity, as mostly derive from the surface during initial setting time for the fact that upper face of mortar are thicker (for the reason that the aggregates yield and thicker grout is pressed on the surface) or are caused by dry shrinkage, those cracks developed in construction will result in structural damage due to erosion in reinforcement (this is called erosion damage), when structures are exposed to air containing sulfur dioxide, carbon dioxide and even small amounts of sulfurated hydrogen, chlorine or other industrial erosive media. Besides, structures with hard-draw high-strength steel bars have more tendencies to erode for the reason that steel bars become more susceptible to erosive media after being cold-drawn when becoming visibly coarse.

g. In some cases, roof structures are seriously overloaded and consequently crack and even collapse. For roofs of foundries, cement mills, and those workshops nearby, without timely cleaning-out of dust, overload incidents are more likely. Other possible overload accidents occur when the roof structures of workshops are rebuilt without precise estimation and design. For example, in 1958, the roof floor of a factory in Handan was designed to substitute 10 cm-thick white slag for 4 cm-thick foam concrete. After a rainy day, the practical load amounted to 1.93 times design value and caused collapse.

h. Violation of construction criteria breeds great accident probability. For instance, roof plates are required to connect with trusses by three spot welds, which often cannot be satisfied in practical operation, and therefore, the bearing capacity of horizontal braces in upper trusses is greatly weakened. For instance, in the process of lifting a composite roof of a workshop, there were many missing welding points between roof plate and truss, which caused weak welding fixation between roof truss and columns and the erected incline, and eventually structural collapse 3 months after construction.

i. Change in internal forces of roof trusses and overwhelming cracks in rod members can be caused by differential settling of foundation. For the case of a factory, located in non-self-weight collapse loess area, the bearing roof structure system was designed as a three-span spandrel-braced consecutive reinforced concrete truss. A great number of cracks (the maximum width was up to 0.7 mm) developing in tensile chords and web members as a result of differential settlement in the basement forced the roof trusses to be reintegrated 2 years after production.

(2) Prestressed concrete roof truss systems

The following problems might easily emerge:

a. Initial eccentricity due to inaccurate location of hollow conduits of prestressing tendons in combination with unbalanced tension of steel tendons at both ends after pretensioning are prone to induce lateral buckling and longitudinal cracks in lower chords.

b. With regard to the relatively large length of lower chords in the roof truss system, flash welding should be adopted to prolong prestressed tendons. Fracture probably happens from poor welding quality. For the 24 m-long span prestressed concrete roof truss system in a certain factory of Nanchang city of China, the lower chords made of 32 mm-diameter, cold-drawn grade II steel bars in this system are jointed using flash welding method. Four years after the workshop was put into operation, a deafening blast was accompanied by the rupture of one prestressing tendon at one side of the lower chord. It was found that the rupture occurred at the welding parts, with over 10 arc craters in the welding cross section, the total area of which represented 15.8% of the entire cross sectional area.

c. As one of the main criteria for the prestressed concrete technique, the quality of tendon anchorage device plays an important role in the anchoring effectiveness. In a factory in Xi'an, the lower chords in the 12 m-long span prestressed concrete bracket are 4 Φ 32 and 2 Φ 25, using bolt bar as the anchorage device. One of the 25 mm-diameter prestressing steels broke suddenly at the joint between screw cap and padding 5 years after the structure was finished. The other end of the broken steel was extended more than 1 m away from the bracket because of the leakiness of grouting. Through chemical component analysis and hardness examination, it was noted that the HRC value of the fractured screw section was 42° to 45°, greatly larger than the design value of 28° to 32°. Another cause of the rupture was uneven loading of the prestressing steels. In a 24 m-long span trapeziform prestressed concrete roof truss of a workshop in Nanjing, fracture was discovered in a great number of joints between bolt bar and prestressing tendons on the morning after tensioning and grouting. The chemical analysis proved that fracture was mainly the result of different chemical components of bolt bar and main prestressing steels that led to poor weldability.

d. If the strength of concrete at automatically anchoring ends does not approach that of the C30 concrete, pre-loosening of steel tendons will probably trigger ineffective anchorage of main reinforcing bars, and even contraction and slippage on the interface, that will eventually result in great stress loss. When alumina cement is utilized to grout into automatically anchoring ends, if the quality of cement cannot be strictly controlled, incipient strength of cement after grouting is unable to reach that of C20, and therefore the main prestressing bars will be out of anchorage, which usually leads to the cracking at edges of roof truss.

e. If prestressing force is exerted to the steel tendons or hyper-tensioning is conducted before concrete strength reaches the design value, it will introduce over-contraction in lower chords, with increase in secondary stress of the other neighboring chords, and even fracture at both ends due to insufficient local bearing pressure capacity. Other cases are longitudinal cracks in lower chords or upper ones for too high pre-compression value.

f. Longitudinal cracks on the surface of hollow conduit are possibly induced by compression exerted from the concrete expansion, when concrete is grouted below 0°C. For instance, certain 36 m-long span prestressing concrete roof truss in Shenyang was under construction in winter. Free water in grouted mortar expanded after a sudden drop in temperature after grouting. And that led to 600 to 1000 mm long cracks on the thinning part of the conduit wall.

2. Problems unique to different types of roof truss systems

(1) Trapeziform roof trusses

a. Difference in reinforcement for different bays is the cause for longitudinal cracks. The method of *different reinforcement for various bays* is often adopted for the sake of saving steel cost, because stresses in upper and lower chords differ dramatically among each bay. In this case, curtailed main reinforcements should have enough anchorage, or cracks will appear in the direction of main reinforcements near the curtailing area.

b. Anchorage of main reinforcements fails at the edge of lower chords. Since the tensile force in lower chords is much greater (in 6 m column-to-column distance and 18 m-span roof trusses, tension in lower chords is around 500 to 600 kN, and for 20 m-span ones, tension is up to 660 to 800 kN), specially designed plates should be placed at both ends of the main reinforcing bars. Cracks will probably occur in both ends of the lower chords in case that those steel plates are neglected or poorly welded. For example, in the 24 m-span trapeziform concrete roof truss located in Zunyi, the lower chords cracked in both ends after erection in 1981. This accident should be ascribed to the missing weld of plates on main reinforcement at the end of lower chords.

c. Poor capacity of anti-cracking and therefore large number of cracks are usually induced by great gap between practical loads and estimated internal forces, which results from great secondary moment at the second web chord from the end.

(2) Composite roof trusses

For the composite roof truss system, joints are so complex to construct that even a little carelessness will cause cracks in those local connections, and eventual collapse of the entire structure. Such an accident happened in the vaulted composite roof truss of an iron shop in Hangzhou. Breakage of joints induced the collapse of the roof trusses. In other regions like Shanxi, Liaoning, Xinjiang, and Henan, there have been similar incidents.

3. Hazard degree analysis for roof truss systems

Evaluation for the deteriorating roof truss systems to quantify the force and stress in the structure is vital when an unsafe condition has been observed. Thus, the hazard level for human safety in the structure greatly increases. Generally speaking, if one of the following described cases happens, timely strengthening becomes a necessity.

a. When overloading exceeds design load value, or practical conditions are beyond the consideration of designers;

b. Concrete strength of actual structure is lower than that of design demand;

c. In case cracks extend along the entire cross section or longitudinal cracks appear in the lower chords in the direction of main reinforcements. Thus, not only effectiveness of bond between steel bars and concrete will be reduced, but also main reinforcements will erode, and concrete coverage will spall;

d. When longitudinal cracks appear at the abutments of both ends, it is concluded that anchorage at bars is unreliable, or the reinforcement and anchoring steel plate are not well jointed. In addition, ineffective anchorage of one tensile web member will cut down its internal force, and simultaneously raise that of the other neighboring ones;

e. In case stiffness of the roof truss system is inadequate, deflection of the structures of this type will often surpass the prescribed limits of deformation demand in accordance with *Criteria of Reliability Assessment for Industrial Buildings* ($> L/400$) ahead of full-loading state;

f. If the roof truss structure without any protection is surrounded with super-high environmental temperature, or extra high level of humidity, or corrosive media, and the width of cracks is over the prescribed value of 0.2 mm;

g. When steel bars become erosive due to carbonization of concrete or some other causes, and covering concrete is exploded;

h. The lack of integrity of roof bracing system causes vibrations and swings of the roof truss when cranes or other machines are at work.

i. Erection sag of the roof trusses cannot meet the criterion requirement and no supporting measures have been adopted.

3.4.2 Retrofit Method for Concrete Roof Trusses

As the interaction of each member in the roof truss system seems so dramatic it is of great importance to choose an appropriate retrofit method and detailing design as well. Hence analysis for internal force in the structure both before and after retrofit can guide method selection and also provide useful information for design. In the following section, key points in internal force analysis will be covered at first, and then some common methods and examples will be discussed.

1. Key points in load calculation and internal force analysis for reinforced concrete roof trusses

(1) Loads and load combinations

The accurate determination of the loads to which a roof truss will be subjected requires consideration for the practical situation, and prediction of the following two load combinations: one is superposition of dead load and entire-span distributed live load exerted, and the other is dead load together with live load distributed in the range of half span. And then the most dangerous combination situation is selected.

(2) Diagram for calculation and computation of internal force

Strictly speaking, entirely cast reinforced concrete roof truss systems are termed as multiple statically indeterminate trusses with restrained connection, and therefore rather complicated calculation procedures are required. But in general they can be simplified as pin-connected trusses. The diagrams for layout and calculation of one mansard roof truss structure are shown respectively in Fig. 3.43(a) and Fig. 3.43(b), where P_a, P_b, P_c,..., P_n correspond to concentrated loads from roof plate, g is defined as the weight of upper chords, and G_1, G_2 and G_3 respectively represents the gravity load of web chords, lower chords and bracings (which are transformed into the joint loads). The load action on the upper chords comprises forces on the joints and between the joints and the latter component brings in flexible deformation and bending moment as well along the upper chords. Although cast together with the upper chords, the web members have little constraint on the bending deformation of the upper chords, considering the greatly decreased stiffness of the web chords, thus, such roof trusses can be modeled as trusses with continuously pin-connected upper chords for simplification, which will account for the scheme used for internal forces in Fig. 3.43(b). The joint loads on the truss equal resistance at the supporting points along consecutive upper chord beams, which can be approximately substituted for pin-connected beams.

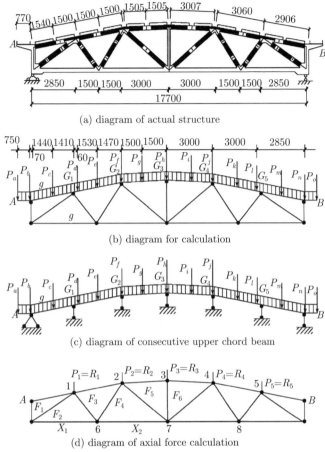

(a) diagram of actual structure

(b) diagram for calculation

(c) diagram of consecutive upper chord beam

(d) diagram of axial force calculation

Fig. 3.43   Calculation diagrams of roof truss.

For the upper chords, internal forces include flexible moments, in addition to axial forces obtained from the pin-connected truss system. The considered moments are figured out using the moment distribution method, in assumption of a mansard consecutive beam with no-freedom pin connections in Fig. 3.34(c).

What should be particularly specified is that the practical loads might deviate from the result obtained through the procedure described above. The main reasons are: ① additional moments in all member bars are derived by relative deflection in the joints under service loads and that those joints are not "ideal hinges"; ② additional moment is caused by non-superposition of resulting force line of reinforcement and external force line, which is due to deviations of reinforcement in construction; ③ if supporting points at two ends of the roof truss are both welded to the top of the column, additional axial force will develop in the structure. Accordingly, appropriate modifications are needed with regard to specific loading situation and detailing design.

2. Retrofit method for reinforced concrete roof trusses

In general, retrofit for reinforced concrete roof trusses can be classified into two types, one is strengthening, which usually applies to cases for local member bars in roof truss, and the other one is removal of load, which fits in with guaranteeing bearing capacity of the global structure. If the roof structure is damaged beyond the capabilities of retrofitting, it should be removed and reconstructed. For further purpose, retrofit can be categorized as follows:

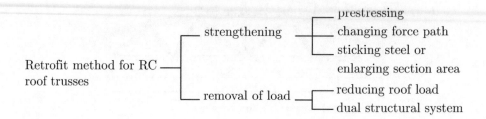

The damage level of the structure and availability of operation should be considered to determine which scheme to adopt from those above. Features and availability for each of these schemes are to be described below.

(1) Prestressing

a. Technique: The method of prestressing is common for its convenience in construction, low cost in material and remarkable effectiveness. More of the lower chords require retrofitting as tensile bar members are more likely to fail. Prestressing reduces the internal force in the tensile chords and raises the bearing capacity, makes crack width narrower, and even closed. In addition, this method can decrease the deflection of roof truss and alleviate the stress lag, thus improving the effectiveness of service.

A variety of arrangements for prestressing tendons are adopted, such as straight-line, sunk, butterfly and composite style.

a) Straight-line style: Fig. 3.44(a) shows how the trapeziform roof trusses in a certain factory of Nanjing has been retrofitted. The problems of this structure are insufficient strength, cracking of concrete lower chords and erosion in steel bars. Though several prestressing methods of retrofitting are discussed, the method of anchoring prestressing tendon at the plate with lobes is adopted. The distance between tendon and lower chord is 250 mm. Prestressing is exerted through tightening reinforced bars with U-shaped bolts at the point 3 m from the anchorage.

b) Sunk style: Fig. 3.44(b) shows the 15 m-long span composite roof truss in a factory. The deficiency in the structure is due to low carrying capacity should be retrofitted in the sunk style. In doing so, the load on the upper chords will be reduced. This not only upgrades the strength of tensile lower chords, but provides reinforced bars with prestressing force, imposing upward force on the truss as well. Prestressing works through electrical heating, in which two ends of the steel bars are welded to the truss after heating when the current is switched on, based on the principle of *thermal expansion and contraction*.

c) Butterfly style: Fig. 3.44(c) shows the trapeziform roof truss in a certain factory of Shanxi, which uses butterfly type retrofit for inadequate strength. Upward force resulting from this type of method is much greater than that from the sunk style, so load reduction is more effective. What is negative is that it might partly transform the mechanical characteristics of some bar members, and even bring about adverse effects. That's why internal force in the truss should be checked to ensure safety of the members. The procedures for this example will be further discussed in the following section.

d) Composite style: Fig. 3.44(d) shows how the roof truss in a steel foundry was retrofitted, where both butterfly type and straight-line prestressing method were adopted. This method employing two different styles simultaneously was called composite style, in which not only lower chords but also tensile web chords can be strengthened.

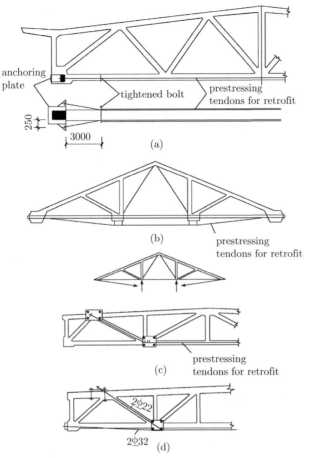

anchoring plate

tightened bolt

prestressing tendons for retrofit

250

3000

(a)

(b)

prestressing tendons for retrofit

prestressing tendons for retrofit

(c)

$2\phi22$

$2\phi32$ (d)

Fig. 3.44   Arrangement of prestressing tendons.

b. Computation of internal force: the process of force computation can be divided into two steps. Firstly, original structural force at service is to be calculated according to the method mentioned in previous section. Dimensions and axial forces of the members of a trapeziform roof truss under external loads are shown in Fig. 3.45(a). Secondly, the internal force under prestressing effect, which is taken as the external force, is to be calculated. Summing up the values obtained in both steps will come to the final result.

Prestressing tendons are placed along 2-10, 10-9, and 9-6 members, and vertical tensions are exerted at points 11 and 8; thus some changes occur in the axial force of this concrete roof truss system after prestressing force has amounted to 100 kN. See Fig. 3.45(b). It can be concluded that each member with prestressing bar decreases by 100 kN in value of internal force, while other ones experience no change. Similarly, if the 11-10, 11-9 and 9-8 members are equipped with prestressing tendons and tensions are put at 11 and 8 joint, internal force reduction will be limited in those members above.

Prestressing force has a wider range of effect when prestressing bars are not arranged along the axis. The value of axial force in the members changes after prestressing tendons are connected to joint 1, 10, 9 and 7, which is in turn loaded 100 kN force. See Fig. 3.45(c). It is evident that axial forces in member 11-2, 11-10, 10-4 and 10-9 are reduced, while those in member 11-1, 1-2, 2-3, 3-4 and 2-10 rise, especially for the 2-10 chord (which should be given more attention).

c. Check calculation of the bearing capacity: this step can be carried out after modification in the resulting internal force of each member with regard to practical load condition. The upper chords are calculated as eccentrically compressive members and the web and lower chords as axial compressive or tensile members in accordance with rules in *Design Criteria for Concrete Structure*.

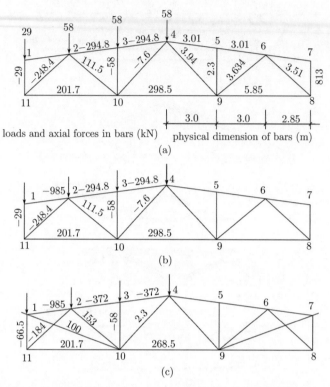

Fig. 3.45   The change in axial forces before and after retrofitting.

(2) Method of sticking steel

a. Technique: strengthening technique of sticking steel is widely used in both tensile and compressive lack of carrying capacity. Much attention should be paid to the quality of anchoring steel angles at tensile bars and the spacing between lacing bars in case of instability in the angle. Two types of sticking steel are provided: dry condition type (see Fig. 3.46(a)) and wet condition type (see Fig. 3.46(b)).

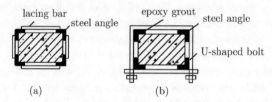

Fig. 3.46   The cross sections of members with sticking steel method.

This method performs especially well in raising the capacity of bar members in roof truss systems, at the cost of minor effect on decreasing the number of cracks in tensile members, especially for dry condition type.

b. Procedure: design for sticking steel angle to retrofitting reinforced concrete roof trusses is summarized as follows:

a) Calculate the internal force under service loads following the scheme described previously;

b) Check the capacity of all members in the structure according to *Design Criteria for Concrete Structure.*

Those deficient members evaluated by the capacity check are to be strengthened by sticking steel. After retrofit, the strengthened members should be analyzed for bearing capacity. For web and lower chords, the method corresponding to checking sticking steel retrofit is applied; for upper chords, a check of prestressing-retrofitted tensile members can be adopted.

(3) Method of changing force path

For upper chords that have weak eccentric compressive capacity, the method of changing the force path satisfies the need. This method is commonly used by setting up inclined struts or subdivisions. Only the former scheme is detailed here.

As presented in Fig. 3.47, an inclined strut acts as one bar to minimize both the span distance between two upper chord joints and eccentric bending moment, with the base of sway rod link to the joint and the top to the joist of the strengthened steel angle. In prevention of failure in supporting due to slip of joist, epoxy-cement mortar or high strength mortar should be laid between the joist and upper chord during construction. And the top end of the joist is tightly propped to the joint with U-shaped bolt. The sway rod can be either welded or bolted with high strength friction grip bolts to the joist. Then, some pre-propping forces will be established by cramming the space between diagonal rod bottom and concrete with steel wedge.

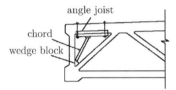

Fig. 3.47   Retrofit method of changing force path.

In doing so, the external moment of global roof truss decreases with increasing numbers of web chords, and the original web bars around additional diagonal web members will be relieved to some extent. Some negative effects will probably develop when those original bars experience adverse changes in mechanical state. It indicates that analysis for internal force in truss members should be repeated after retrofitting.

In the process of internal force computation, the additional web member serves as the compressive web bar, following the previously discussed method. If the number of additional web members is small or they are only required to be placed in inter-joints at two ends of roof, the members in the specific region are required to be recalculated.

(4) Method of reducing roof load

Through reducing the loads on the roof, the force in the members of the roof truss structure will be curtailed accordingly, which is an effective approach to ensuring the roof truss safe, solving such issues as easily cracking and insufficient bearing capacity. This method is fulfilled by modifying the structure type of roof truss and reducing weight of roof. Replacing the large-scale roof floor with corrugated iron or asbestos cement tile, and water-proof with light-weight lamina are good examples.

(5) Method of dual structural systems

Dual structural system method refers to adding a new roof truss system up to the original one, to sustain loads. One of the techniques is to build a new truss across the original two bars of truss, and this will relieve the burden of the original structure and also reduce the spacing of frame and span of roof plate. This might bring about some difficulties such as how to properly place the new structure and prevent hogging moment at the middle supporting point of the roof floor, where the new one is located. Another scheme is binding one bay of roof truss that is usually made of steel or lightweight steel at each side of the original structure.

For such reasons like difficulty in construction, high cost of steel material and unsatisfactory compatibility between new and original structures, the proposed method is rarely used.

3. Measures of improving durability of concrete roof truss systems

Insufficient durability often leads to serious erosion in members and spalling of concrete covering induced by carbonization of concrete in tensile members or over-wide cracks, and decreased safety of the roof truss structure.

Retrofit for durability should include two types of measures, one to prevent or retard the steel bars from further erosion; and the other to strengthen seriously eroded tensile members.

The approaches to preventing deterioration are detailed in other related materials. Furthermore, three steps of strengthening are reasonable for a small area of cross section of members: first, seal the cracks with waterproof airtight material; next, cover the surface with waterproof paint; and wrap the member using the "one-cloth, two-glue" method, that is, brushing epoxy resin while binding the gauze, and a second layer of epoxy resin follows. Such method is suitable when cracks are developed in one side and the reinforced bars are located near that side.

For eroded lower chords, the method of adding tensile bars is recommended on account that the high-strength reinforcement which is commonly arranged in the roof truss structure is prone to fracture after having pitting, and the lower chords become crucial to the performance of the entire roof. Thus, substitute for original members not only enables the strengthening bars perform well in carrying tension, but also prevents the roof truss from collapse even when some of the steel bars in the lower chords rupture. Strengthening is fulfilled through the technique of prestressing.

3.4.3   Practical Examples of Retrofitting of Concrete Roof Trusses

Example 3.10   A certain steel-casting foundry is 108 m in length and 18 m in span, which is divided into eastern and western parts by the dilatation joint. The eastern part is for electric furnace smelting and the western part is the area of molding and casting. The roof truss system was designed as the ΦTM-18 trapeziform reinforced concrete structure by the former institute of design for iron and steel industry. The construction was finished in 1959 and concrete cracked before use due to poor construction quality. Consequently, the lower chords were simply strengthened by means of adding ϕ30 reinforced bars at both ends and fixing the top of truss with nuts. In 1967, it was found that there were a total of 57 cracks on 16 bays of the truss, the widest one measured 0.5 mm and that those cracks mainly distributed on two sides of the lower chord joints and diagonal web chords. These problems fostered the second retrofit that included replacing the reinforced bars in lower chords with $2\Phi32$ prestressing bars in sunk style and adding $2\Phi22$ bars to both sides of diagonal web chords accompanied by screw-induced tension. See Fig. 3.48.

After 10 years, the examination concerning the 16 retrofitted trusses indicated that a great number of cracks appeared, especially in the region of lower chords, where a large number of cracks initiated from the tops of chords. The number of cracks added up to over 300, of which 56 were at both sides of the lower chord joints, 145 lay along the lower chords, and 105 were in the inclined web chords. Those cracks more than 0.3 mm in width totaled 28, the widest of which equaled 0.6 mm. This structure was assessed as a "structure at risk" in accordance with standard. In 1980, it was decided to conduct the third retrofit design and construction, which had been determined by internal force analysis.

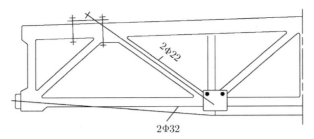

Fig. 3.48   Roof truss with retrofit of composite style.

(1) Analysis for causes of cracking

After the second retrofitting, the cracks on the original members developed instead of having been controlled, especially on the lower chords, where those cracks began from the upper regions. The causes for the irregular phenomenon are listed in the following:

a. The cross-section areas of the lower chords were relatively small (only 200 mm × 200 mm), and reinforced bars were arranged in the form of a line, concentrating at the axial lines. The top and bottom area of cross section was plain concrete, so the tensile force in that area would be totally carried by concrete only under eccentrically tensile loads, which would result in great development of cracks once cracking initiated.

b. The cracks after the second retrofitting of sunk style did not decrease but increased in number. This indicates that the line of tension in prestressing tendons deviates from the axial line; the prestressing tendons kink at the point of downdip, forming the upward force, not through the joints of members. The friction between steel channels at the point of downdip and prestressing tendons restrains the tendons from slipping and causes the difference in tensile forces which derive from unequal tensile stresses in the tendons at two sides of the downdip point. In that way the additional tensile difference adds the bending moment value in the concrete members and consequently both tension in the top region and compression in the bottom region or relatively greater tensile stress on the top make cracking begin from the top. And the full-size experimental results have proved it.

The analysis above shows that the members with line-laid reinforcement behave more sensitively to eccentric force effects and it can be concluded that it is better for the lower chords with line-laid reinforcement to adopt straight-line style retrofit method when using prestressing bars; if sunk style is used, the sleek supporting point will permit prestressing bars slipping without restraint. In addition, it should be noted that the reaction in the upward direction is to pass the joints of all the members.

(2) Decision-making and full-size test

This building was proposed to use the retrofit method of dual structural system, that is, a pair of additional steel roof trusses are constructed on both sides of the original structure, with additional structure supporting on the steel columns for the reason that this building has been defined as "structure in risk". Cracking in the lower chords was severe and irregular and the eastern part was subject to fire accident. Despite reliability, this method is deficient in economy and construction is difficult. Both theoretical analysis and full-size experiment should be conducted for the original structure for the sake of safety, economy and rationality.

The result after checks shows that the upper chords roughly satisfy the requirements except those in both end bays are dramatically insufficient in both bearing and anti-cracking capacity considering all the effects of axial force, bending moment and secondary moment. The bearing capacity of lower chords is merely 80% of the design value when neglecting the effects of additional prestressing bars; if the additional prestressing force is considered, both of the bearing and anti-cracking capacity will meet the criterion. Except for the fact that the first diagonal one lacks in bearing capacity, most of the original web members measure up. If the additional prestressing effects are included, that deficient member becomes satisfactory.

In the full-size experiment of the original structure, resistance strain gauges are arranged on the upper and lower chords, diagonal tensile members, additional prestressing tendons and tensile reinforced bars to test its load situation. The result reports that the internal forces in those tested members become different to some larger or small extent as the roof loads change, which proves the effectiveness of the second retrofit design since the original and additional members are able to work together compatibly.

(3) Retrofit program

On the basis of computation, analysis and test, the method of binding steel roof truss dual structure is chosen, considering the high temperature of electric furnace smelting in the eastern part. For the western part, reinforcement on the original retrofit follows these three points:

a. Adding sway rods and altering the load path can solve the problem of insufficient bearing capacity of the upper chords in both end bays, such as setting up joists with angles and stay bars in that area (see Fig. 3.48) and strengthening vertical bars in the bottom with a formed steel jacket.

b. Although the lower chords crack seriously, the bearing capacity is adequate and the additional prestressing tendons in sunk style method can work together well with the original members despite the inappropriately designed detailing. Therefore, blocking some concrete of C30 grade into the steel channel anchorage at the prestressing tendons and then tightening the anchoring caps on them are enough to prevent deformation in anchoring plate from having negative effects on the load state, with no retrofit in the middles of lower chords.

c. It is found that the upper chords in the second bay are unsafe because of the additional diagonal stay bars in the first bay, which require the jacketing formed steel to strengthen them.

d. For other tensile web bars, the screw caps on the original prestressing bars should merely be tightened.

After the retrofit construction mentioned above, the roof truss system performed well.

CHAPTER 4

Retrofitting Design of Masonry Structures

4.1 Introduction

Masonry is a kind of load-bearing material made of the block materials (bricks or masonry blocks) bonded together by mortar. Compared with concrete, masonry has a certain capacity of compressive strength but lower capacity in tensile, shearing and bending strength. Therefore, masonry structure especially plain masonry has a very poor integrity, rather low capacity of load-bearing and is vulnerable to crack under external loads. Masonry needs to be repaired and retrofitted in the following situations:

a. Settlement cracks were formed in the wall because of the uneven settlement of foundation.

b. Temperature cracks were engendered in the wall because of thermal expansion and contraction of roof.

c. Bearing capacity was insufficient in local masonry walls and columns.

d. Deficiency of bearing capacity in original masonry structure caused by adding stories or rebuilding.

e. Seismic strength is insufficient or unsatisfactory and seismic constructional measures of the building are not satisfactory according to the earthquake resistant evaluation in seismic protection area.

f. Building damaged after earthquake.

Retrofitting methods in common use for masonry structures include the methods of direct retrofitting, changing load transferring path and adding outside structure.

Direct retrofitting method refers to retrofitting or repairing the structural member whose strength is insufficient and constructional measures cannot meet the demand, without changing load-bearing system and plan layout.

The method of changing load transferring path means changing plan layout and load transferring path. It usually necessitates adding bearing walls, columns, and their corresponding foundations.

The method of adding outside structure means adding concrete or steel structure outside the original structure to transfer partial loads of original structure and loads of extension story structure directly to the foundation through outside structure. It is mainly used in the construction of additional stories.

This chapter mainly introduces several usual direct retrofitting methods for multi-story masonry brick buildings. The retrofitting design of multi-story block masonry buildings, inner-frame masonry buildings and frame supported masonry buildings are covered in Chapters 3 and 4.

4.2 Repairing and Strengthening of Wall Cracks

Wall cracks appear when uneven settlement, thermal expansion and cold shrinkage, lack of bearing capacity and earthquake occur. Repairing and strengthening may be used for crack retrofitting when crack expansion stops.

a. Grouting can be used in large walls with few cracks.

b. When walls crack so seriously that grouting is unfeasible, tearing down and rebuilding may be considered and in this case it is better to use pressure grouting in the closed cracks.

Materials such as pure cement slurry, cement mortar, sodium silicate mortar, and cement lime mortar and so on can be used for grouting (Table 4.1). Pure cement slurry might be a good choice for masonry repairing for its excellent groutability. It will be easily injected into surface-penetrating pores and 3 mm-wide cracks can be compactly grouted. Cement mortar can be adopted when the width of crack exceeds 5 mm, while pressure grouting may be used for fine cracks. The row of slurry in Table 4.1 is suitable for cracks of 0.3 mm to 1 mm; thick slurry can be applied to cracks of 1 mm to 5 mm and mortar for cracks more than 5 mm.

Table 4.1   Reference mixture ratio for crack grouting materials

Categories	Cement	Water	Binding material	Sand
Slurry	1	0.9	0.2 (107 Adhesive)	
	1	0.9	0.2 (Binary Emulsion)	
	1	0.9	0.01~0.02 (Sodium Silicate)	
	1	1.2	0.06 (Polyvinyl Acetate)	
Thick	1	0.6	0.2 (107 Adhesive)	
slurry	1	0.6	0.15 (Binary Emulsion)	
	1	0.7	0.01~0.02 (Sodium Silicate)	
	1	0.74	0.055 (Polyvinyl Acetate)	
Mortar	1	0.6	0.2 (107 Adhesive)	1
	1	0.6~0.7	0.15 (Binary Emulsion)	1
	1	0.6	0.01 (Sodium Silicate)	1
	1	0.4~0.7	0.06 (Polyvinyl Acetate)	1

Sodium silicate mortar with sodium fluorosilicate can be used for wide cracks; its mixture ratio of sodium silicate, slag powder and sand is (1.15~1.5):1:2, then 15% sodium fluorosilicate with 90% purity is added.

Take pure cement slurry for example; its construction procedure is shown as follows:

Step 1, clear the crack and make sure it is unblocked.

Step 2, grout the crack by 1:2 (cement to water) cement mortar with accelerator to avoid slurry leakage when grouting.

Step 3, grouting hole or mouth should be made near the top of the crack with electric drill or hand hammer.

Step 4, flush the crack by 1:10 (cement to water) slurry and check the patency of crack, meanwhile moisten masonry around the crack.

Step 5, grout the crack with pure cement slurry (the ratio of cement to water is 3:7 or 2:8).

Step 6, allow local curing at the region of grouted crack.

Construction of pressure grouting is similar to the above, but first check the extent of passage leakage by compressed air under the pressure of 0.2~0.25 MPa. The leakage must be plugged if too severe.

For horizontal full-length cracks, holes can be drilled and pin keys formed to strengthen the interaction of two sides of the wall. The diameter of pin keys is 25 mm, spacing is 250~300 mm and depth can be 20~25 mm thinner than the wall. Then grout after finishing pin keys.

c. The retrofitting method of adding local cement mortar layers with reinforcement mat is available when the wall cracks are densely distributed.

For a cracked wall lacking bearing capacity, methods of increasing the load-bearing capacity should be adopted while repairing cracks.

4.3   Retrofitting of the Wall: Lack of Load-bearing Capacity

4.3.1   Retrofit of Brick Wall with Buttress Columns

Using buttress columns is the most common method for wall retrofitting and it can increase equivalent wall thickness and wall section area effectively or reduce effective height of the wall; thus compressive bearing capacity will be enhanced effectively. There are two kinds of buttress columns, i.e. brick buttress and concrete buttress columns.

1. Constructional measure for brick buttress column

Conventional brick buttress columns are shown in Fig. 4.1, in which (a) and (b) represent one-side brick buttress columns while (c) and (d) represent two-side brick buttress columns.

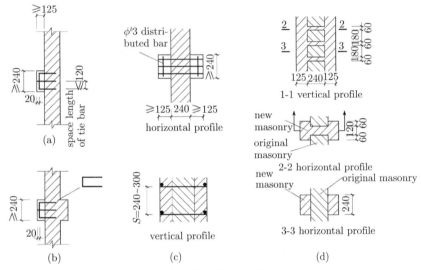

Fig. 4.1   Retrofit of brick wall by brick buttress column.

Connection of added buttress columns and original brick wall can be made by inserting reinforcements or digging and inlaying to ensure their interaction.

Fig. 4.1 shows the connections of inserting reinforcements. Specific methods are as follows:

a. Shuck off the plaster layer of contact surface between new and original masonry, and flush it clean.

b. Implant connection dowel rebars of ϕ^b4 mm or $\phi6$ mm in the mortar joint of brick wall; drill holes with electrical drill before implant connection rebars if it is difficult to implant. Horizontal distance of dowel rebars should not be larger than 120 mm (Fig. 4.1), and vertical distance is 240~300 mm (Fig. 4.1).

c. Band closed rebar of ϕ^b3 mm along the open side (Fig. 4.1(c)).

d. Build buttress columns with mixed mortars of M5~M10 and bricks of more than MU7.5. Width of the columns should not be less than 240 mm and thickness not less than 125 m. When built to the bottom of floor or beam, use hardwood top bracing or build the last five layers of vertical mortar joint with expansive cement to ensure strengthening masonry functions effectively.

Fig. 4.1(d) shows the connection of digging and inlaying. Specific procedure is to remove roof bricks of the wall and enchase cast-in bricks when building bilateral buttress columns. It is better to incorporate appropriate expansive cement in the mortar for cast-in bricks in the original wall to guarantee the tightness of cast-in bricks and original wall.

The distance and quantity of buttress columns that the brick wall demands should be determined by calculation.

2. Bearing-capacity checking for retrofitted wall with brick buttress columns

Considering stress-lagging in post-build buttress columns, factored value f_1 of compression strength of buttress columns should be multiplied by discount coefficient 0.9 when calculating the bearing capacity of a retrofitted brick wall. Compressive bearing capacity can be checked by the following formula:

$$N \leqslant \varphi(fA + 0.9f_1A_1) \tag{4.1}$$

where N is the axial force induced by load factored value; φ is the influence coefficient for load-bearing capacity of compressive members engendered by depth-thickness ratio β and eccentricity of axial force e, and its value can be obtained from *Code for Design of Masonry Structures*; f and f_1 are design values of compression strength for original wall and buttress columns respectively; A and A_1 are the areas of original wall and buttress columns.

It is not necessary to consider the stress-lagging of buttress columns when checking depth-thickness ratio of a retrofitted wall and requirements for serviceability limit state. Combination section after adding buttress columns should be adopted to calculate influence coefficient for bearing capacity of compression members.

Example 4.1   An office building has a cross wall thickness of 240 mm, a spacing of 4 m between the cross walls, depth of 6 m, story height of 3 m, and the reinforced concrete floor thickness of 120 mm. The pressure transverse walls bear is obtained by calculation, and its value is 188 kN/m. Site inspection results showed that the strength grade of bricks is about MU7.5 and the mortar is M0.4.

a. Check the original brick wall capacity. Design value of masonry compression strength in masonry structure: $f = 0.79\text{MPa}$ of brick compression strength can be obtained from *Code for Design of Masonry Structures*.

The calculated height of the wall H_0: According to rigidity plans, because of $s = 2H$, where s is the spacing of the walls or pilasters which could act as lateral bracing of the analyzed walls, here it refers to the depth of house. By consulting code list 4.1.3, we obtained $H_0 = 0.4\,s + 0.2\,H = 0.4 \times 6 + 0.2 \times 3 = 3.0$ m.

$$\beta = \frac{H_0}{h} = \frac{3000}{240} = 12.5 < [\beta]$$

$\varphi = 0.59$ can be obtained by code appendix list 5-4, and the design value N_0 for bearing capacity of original brick wall according to the following formula;

$$N_0 = \varphi f A = 0.59 \times 0.79 \times 240 \times 1000 = 111.8 \text{ kN} < N = 188 \text{ kN}$$

Result indicates that the brick wall must be retrofitted.

b. Retrofitting design. Buttress columns are built with bricks of MU10 and mixed mortar of M10. Compression strength $f_1 = 1.99$ MPa by consulting code. Establish buttress columns every 1.5 m on both sides of the original wall. The width of the buttress column of one side is 240 mm, and the thickness is 125 mm (thickness direction is along the depth of wall), as shown in Fig. 4.2.

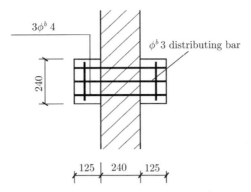

Fig. 4.2   Wall retrofitted by buttress column.

$$I = \frac{1}{12} \times [(150 - 24) \times 24^3 + 24 \times 49^3] = 3.80 \times 10^5 \text{ cm}^4$$

$$A = 150 \times 24 + 24 \times 25 = 4200 \text{ cm}^2$$

The equivalent thickness is

$$h_T = 3.5i = 3.5\sqrt{\frac{I}{A}} = 33.3 \text{ cm}$$

$$\lambda = \frac{H_0}{h_T} = \frac{3000}{33.3} = 9.0$$

Looking up the table, $\varphi = 0.735$ is obtained. According to equation (4.1),

$$N_P = \varphi(fA + 0.9f_1 A_1)$$

$$= 0.735 \times (0.79 \times 240 \times 1500 + 0.9 \times 1.99 \times 240 \times 250)$$

$$= 288 \text{ kN} > 1.5 \times 188 = 282 \text{ kN}$$

3. Technology and construction for concrete buttress column

Concrete buttress column, which is shown as Fig. 4.3, can help the original wall to sustain more loads.

Connection of concrete buttress column and original wall is very significant. For a wall with pilasters, connection of new and original columns shown as Fig. 4.3(a) is the same with brick buttress columns. When thickness of original wall is less than 240 mm, U-shape stiffeners should penetrate wall body and be bent (Fig. 4.3(b)). Retrofitted form shown in Fig. 4.3(c) and(e) can enhance bearing capacity of original wall effectively. As shown in Fig. 4.3(a), (b), (c), vertical spacing of U-shape stirrups should not exceed 240 mm, and the diameter of longitudinal rebar may not be less than 12 mm. Fig. 4.3(d) and (e) display the connection of pin key, the vertical spacing of which should not exceed 1000 mm.

Concrete buttress column usually adopts concrete of C15~C20, and its width and thickness may be not less than 250 mm and 70 mm respectively.

Fig. 4.4 gives the method of strengthening brick wall and pilaster with concrete. Thickness of re-pouring concrete may not be less than 50 mm, and it is better to use spray method for construction. To reduce site work, two open hoops and one closed hoop can be placed alternately for retrofitting of original brick wall and pilaster shown in Fig. 4.4(a). Open hoops should be inserted into brick joint of original wall and insertion depth should not be less than 120 mm. Closed hoops should not be bent until they penetrate wall body. Drill holes with electrical drill before inserting loops if the insertion is difficult. Diameter of longitude reinforcement should not be less than 8 mm.

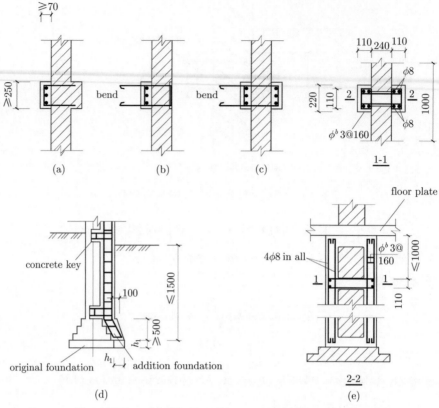

Fig. 4.3   Retrofit of brick wall by concrete buttress column.

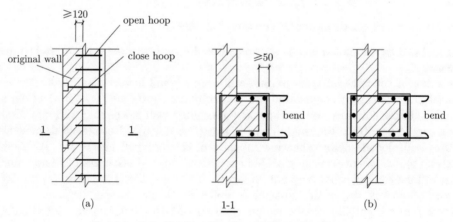

Fig. 4.4   Retrofit of brick wall and pilaster by concrete.

4. Bearing-capacity checking for retrofitted wall with concrete buttress columns

Masonry after retrofitting by concrete buttress columns becomes composite masonry. Considering that post-build concrete buttress columns are related with stress state of original wall and existing stress lagging, strength reduction factor α should be introduced for post-build concrete buttress columns when calculating bearing capacity of composite brick masonry.

Axial compressive capacity of composite brick masonry can be attained by the following formula:

$$N \leqslant \varphi_{com}[fA + \alpha(f_cA_c + \eta_s f_y' A_s')] \tag{4.2}$$

where φ_{com} is stability factor for composite brick masonry, obtained by *Code for Design of Masonry Structures*; α is strength reduction factor for post-build concrete buttress columns. If original brick masonry is in a good condition when retrofitting, $\alpha = 0.95$ can be used; if the masonry has stress cracks and damage, $\alpha = 0.9$ should be adopted. A is the section area of original masonry; f_c is the design value of axial compressive strength for concrete buttress columns or mortar layers. The design value of axial compressive strength for mortar can adopt 70% concrete design value of the same strength grade, it is 3 MPa when mortar is M7.5; A_c is section area of concrete or mortar layer; η_s is strength coefficient of compression reinforcement, its value will be 1.0 for concrete layer and 0.9 for mortar layer; A_s' and f_y' are section area and design value of compressive strength of compressive reinforcements respectively.

Bearing capacity of eccentric compressive composite brick masonry (Fig. 4.5) can be calculated as the following formulas:

$$N \leqslant fA' + \alpha(f_cA_c' + \eta_s f_y' A_s') - \sigma_s A_s \tag{4.3}$$

or

$$Ne_N \leqslant fS_s + \alpha[f_cS_{c,s} + \eta_s f_y' A_s'(h_0 - a)] \tag{4.4}$$

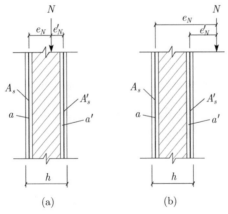

Fig. 4.5   Eccentric compression member of composite brick masonry.

Here the height of compressive region x is determined by the formula:

$$fS_N + \alpha(f_cS_{c,N} + \eta_s f_y' A_s' e_N') - \sigma_s A_s e_N = 0 \tag{4.5}$$

where A' is compressive area of original brick masonry; A_c' is compressive area of concrete or mortar layer; S_s is the moment of compressive area of brick masonry to gravity center of tensile reinforcements; $S_{c,s}$ is the moment of compressive area of concrete or mortar layer to gravity center of tensile reinforcements A_s; S_N is the moment of compressive area of brick masonry to action position of axial force N; $S_{c,N}$ is the moment of compressive area of concrete or mortar layer to action position of axial force N; e_N' and e_N are distances from gravity centers of compressive reinforcements and tensile reinforcements to action position of axial force N respectively (Fig. 4.5); their values are based on the following formulas:

$$e'_N = e + e_i - \left(\frac{h}{2} - a'\right)$$

$$e_N = e + e_i + \left(\frac{h}{2} - a\right)$$

where e is original eccentricity distance of axial force calculated by load standard value, $e = 0.05h$ will be adopted if $e < 0.05h$; e_i is additional eccentricity distance of composite brick masonry members under axial force, and it can be calculated as:

$$e_i = \frac{\beta^2 h}{2200}(1 - 0.022\beta)$$

h_0 is effective height for section of composite brick masonry member and shall be calculated as $h_0 = h - a$; σ_s is the stress of tensile reinforcements; for large eccentric compression it shall be calculated as $\sigma_s = f_y$ and for small eccentric compression ($\xi \geqslant \xi_b$), it can be calculated as the following formula:

$$\sigma_s = 650 - 800\xi \tag{4.6}$$

a' and a are the distances from gravity centers of compressive reinforcements and tensile reinforcements to closer edge of the section; ξ is relative height of section compressive region of composite brick masonry, calculated as $\xi = x/h_0$; ξ_b is the limited maximum value for ξ, it is 0.55 for grade I reinforcing bar and 0.425 for grade II reinforcing bar.

4.3.2 Retrofitting Brick Wall by Adding Mortar or Concrete Layer with Mat Reinforcement

It is better to retrofit by adding mat reinforcement layer when a brick wall has a serious lack of compressive bearing capacity or shear strength and lateral stiffness are insufficient, but this method is not suitable for the following situations:

a. Aperture diameter of brick is more than 15 mm in hollow brick wall.

b. Mortar strength grade of brick wall is less than M0.4.

c. Wall surface is seriously damaged, or oil residue cannot be cleaned out so that bonding quality of layer and wall cannot be guaranteed.

1. Construction requirement

Retrofitted layer may adopt:

a. Adding mortar layers with reinforcement mats on two sides (or one side).

b. Adding fine aggregate concrete layers with reinforcement mats on two sides (or one side).

c. Adding cement mortar layers with reinforcement mats on two sides.

When retrofitted by the above three methods, the following measures on construction should be taken to ensure that cement mortar or concrete has a reliable bond with original masonry: first remove plastering layer of original wall and pick out 10 mm-depth mortar of brick joints, then clean wall surface by steel brush and moisten it by sprinkling with water. When using fine aggregate concrete with reinforcement net to strengthen wall, 50 mm is suitable for the depth of layer, concrete strength grade should not be less than C20 and diameter of aggregate should not exceed 8 mm, and injection should better be chosen for construction; when using cement mortar with reinforcement net to strengthen wall body, the depth of mortar can be 30~40 mm and mortar strength grade should not be less than M10. Reinforcement net should be tied by rebar $\phi6$ every 1000~1200 mm (Fig. 4.6 (a) and (b)). When wall is retrofitted on one side, chisel hole with section of 60 mm × 60 mm and depth of 120~180 mm on it, then clean it and pre-embed $\phi6$ S-shape reinforcement to tie mat reinforcement (shown in Fig. 4.6(c)), or fix wall by $\phi 4$ mm U-shape reinforcement nailing into wall instead of S-shape reinforcement. To intensify fixation of mat reinforcement and wall, if necessary, $\phi 4$ mm U-shape reinforcement may be added in the middle or iron nails may be nailed into brisk gaps on the wall. When mat reinforcement goes across slab, chisel hole every 600 mm on slab and put $\phi 12$ reinforcement into the holes, then grout them with concrete (Fig. 4.7). Longitude reinforcement should be extended to the depth of 400 mm below indoor or outdoor ground, and be affixed by C15 concrete (Fig. 4.8).

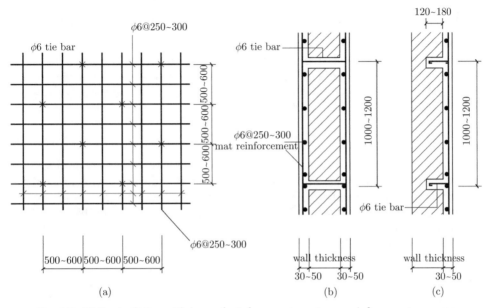

Fig. 4.6   Wall retrofitting with layer of reinforcement mortar or reinforcement concrete.

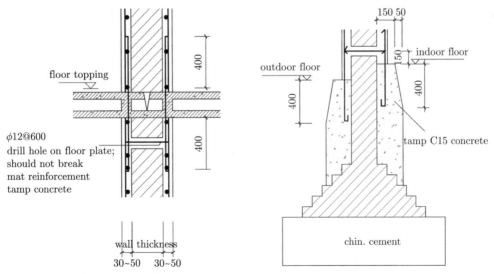

Fig. 4.7   Reinforcement connection at slab.   Fig. 4.8   Reinforcement anchorage in ground.

When retrofitted by cement mortar layers, 20~30 mm is suitable for the thickness of them and mortar strength grade should not be less than M10.

When transverse reinforcement goes across doors or windows, it is better to bend the rebar perpendicular to the wall surface into a straight hook along edges of holes and then anchor it. Holes with penetrating S-shape reinforcement in the wall must be drilled by machine. It is better to drill holes with penetrating reinforcement in a slab by machine.

2. Compressive capacity calculation for retrofitted wall with reinforcement mat layer

Wall body retrofitted with reinforcement mat layer becomes composite masonry and its compressive capacity of normal section can be calculated by Eqs. (4.2)~(4.5).

3. Shear capacity calculation for retrofitted wall with reinforcement mat layer

There are many factors influencing shear capacity for retrofitted walls with layers, such as compressive stress of upper wall, thickness and shear strength of mortar layer, reinforcement layout quantity and strength of layer, thickness and shear strength of the original wall. In reference to relative test results, shear capacity for a retrofitted wall with reinforcement mat layer can be checked by the following formula:

$$V_k \leqslant \frac{(f_{vz} + 0.7\sigma_0)A_k}{1.9} \tag{4.7}$$

where V_k is the shear of the k th wall; σ_0 is average compressive stress of wall section in half of story height; A_k is section area of the kth wall in half of story height (area of doors and windows should be deduced); f_{vz} is equivalent shear strength of original brick wall after retrofitting (conversion shear strength for short) and lower value based on the two following formulas should be adopted according to different repairing and retrofitting conditions, when controlled by layer mortar strength:

$$f_{vz} = \frac{nt_1}{t_m}f_{v1} + \frac{2}{3}f_v + \frac{0.03nA_{sv1}}{\sqrt{s}\,t_m}f_y \tag{4.8}$$

when controlled by reinforcement strength:

$$f_{vz} = \frac{0.4nt_1}{t_m}f_{v1} + 0.26f_v + \frac{0.03nA_{sv1}}{\sqrt{s}\,t_m}f_y \tag{4.9}$$

where t_1 is thickness (mm) of cement mortar layer or mortar layer with reinforcement mat; t_m is thickness (mm) of original brick wall; n is the number of retrofitted layers of one wall; f_{v1} is mortar shear strength (N/mm^2) of layer and can be calculated by $f_{v1} = 1.4\sqrt{M}$, M is mortar strength grade of layer; f_v is design value of brick masonry shear strength along full-length cracks, $f_v = 0$ will be adopted for cracking wall without repairing cracks; A_{sv1} is section area of single reinforcement; f_y is design value of reinforcement strength (MPa); s is spacing (mm) of reinforcements in reinforcement mat, and the unit of $\sqrt{s}$ in the formula is still mm.

Example 4.2   Consider a four-floor brick house with walls between windows. The wall width is 2.7 m and thickness 0.24 m. Its brick strength grade is MU7.5 and mortar strength grade M2.5. Average compressive stress of the wall in half height of the first floor is 0.39 MPa. Design value of seismic shear force the wall is $V_k = 180$ kN. Determine seismic strength of the wall; if its strength is deficient, retrofit the wall by cement mortar layer with reinforcement mat and check again.

a. Check seismic strength of original wall

$f_v = 0.09$ MPa can be obtained as wall shear strength along full-length cracks in *Code for Design of Masonry Structures*. $\xi_N = 1.426$ can be got as normal stress effect coefficient of masonry strength in *Code for Seismic Design of Buildings*. Masonry seismic shear strength is

$$f_{vE} = \xi_N f_v = 1.426 \times 0.09 = 0.128 \text{ MPa}$$

$$f_{vE} \cdot A/\gamma_{RE} = 0.128 \times 2700 \times 240/1.0 = 82944 \text{ N} < V_k$$

where γ_{RE} is 1.0 for a beam on the wall between windows.

b. Checking seismic strength of the wall after retrofitting

Retrofit the wall by adding cement mortar layer with reinforcement mats in one side. The mat is composed of bi-directional $\phi6@300$, thickness of the layer is 30 mm and mortar strength grade is M7.5. Mortar shear strength of layer is

$$f_{v1} = 1.4\sqrt{7.5} = 3.83 \text{ MPa}$$

When controlled by layer mortar strength:

$$\begin{aligned}
f_{vz} &= \frac{nt_1}{t_m}f_{v1} + \frac{2}{3}f_v + \frac{0.03nA_{sv1}}{\sqrt{s}\,t_m}f_y \\
&= \frac{30}{240} \times 3.83 + \frac{2}{3} \times 0.09 + \frac{0.03 \times 1 \times 28.26}{\sqrt{300} \times 240} \times 210 \\
&= 0.478 + 0.04 + 0.06 \times 1.427 = 0.563 \text{ MPa}
\end{aligned}$$

When controlled by reinforcement strength:

$$\begin{aligned}
f_{vz} &= \frac{0.4nt_1}{t_m}f_{v1} + 0.26f_v + \frac{0.35nA_{sv1}}{\sqrt{s}\,t_m}f_y \\
&= 0.4 \times 0.478 + 0.26 \times 0.09 + 0.35 \times 1.427 = 0.714 \text{ MPa}
\end{aligned}$$

where $f_{vz} = 0.563$ MPa is used.

Seismic capacity of brick wall after retrofitting is

$$\begin{aligned}
\frac{(f_{vz} + 0.7\sigma_0)A_k}{1.9} &= \frac{0.563 + 0.7 \times 0.39}{1.9} \times 2700 \times 240 \\
&= 285120 \text{ N} = 285.1 \text{ kN} > V_k
\end{aligned}$$

Design value of seismic shear force of retrofitted wall is $V_k = 216$ kN.

4.4   Retrofitting of Brick Columns for Bearing Capacity Deficiency

Those brick columns with inadequate bearing capacity can be retrofitted by augmenting their sections or utilizing outer packing rolled angles.

4.4.1   Augmenting Sections for Retrofitting Brick Columns

Augmenting section methods for retrofitting brick columns can be done in two ways. One is covering column with a lateral surface concrete layer (called lateral retrofitting, for short). And the other way is packing concrete covers or mat reinforced cement mortar covers around a column, as in Fig. 4.9.

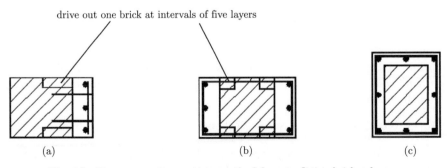

Fig. 4.9   Two augmenting sections method for retrofitting brick columns.

1. Lateral retrofitting brick columns with concrete layer

A brick column under large moment is usually retrofitted by covering a lateral surface concrete layer only on the compression side or on two sides as in Fig. 4.9.

The connection and joint of new and original columns are important when using lateral retrofitting. The connected stirrups should be used in double-sided retrofitting; and when using single-sided retrofitting, concrete nails or expansion bolts should be nailed into the original column to reinforce the connection. Furthermore, with single-sided or double-sided retrofitting, one angle brick of original column should be driven out at intervals of 5 layers, to make a better connection between new concrete cover and original column, as in Fig. 4.9.

A proper strength grade of cast-in-situ concrete may be C15 or C20; the spacing between brick column and bearing reinforcement should not be less than 50 mm; The ratio of compressive reinforcement may not be less than 0.2%, and its diameter should not be less than 8 mm.

After being laterally retrofitted, brick columns become composite brick masonry, whose compressive load-bearing capacity can be calculated by the Eqs. (4.2)~(4.5).

2. Packing all-around concrete covers for retrofitting brick columns

The method of packing concrete all around has a good effect, especially for bearing capacity of brick columns under axial or small eccentricity compression.

Cement mortar may be used in outer cover when the cover is a little thin. Strength grade of cement mortar should not be less than M7.5. Moreover, $\phi 4 \sim \phi 6$ closed stirrups should be set in outer cover and their spacing may not be larger than 150 mm.

Because of closed stirrups, the lateral deformation of brick columns will be constrained. Thus, the behavior of the columns with all-around concrete covers is similar to that of reticular reinforcement brick masonry. And its bearing capacity under compression can be calculated by Eq. (4.10):

$$N \leqslant N_1 + 2\alpha_1 \varphi_n \frac{\rho_v f_y}{100} \left(1 - \frac{2e}{y}\right) A \qquad (4.10)$$

where N_1, bearing capacity of retrofitted combination brick masonry under compression, can be calculated from Eqs. (4.2)~(4.5);

φ_n, influence coefficient of the compressive bearing capacity of reticular reinforcement brick masonry, when considering ratio of height to thickness of reinforcement and axial eccentricity should be determined from *Code for Design of Masonry Structures* (GB50003—2001);

ρ_v, volume ratio of reinforcement (%), if length of stirrup is a, width is b, spacing is s, and section area of single-limb stirrup is A_{sv1};

$$\rho_v = \frac{2A_{sv1}(a+b)}{abs} \times 100$$

f_y, design value of stirrup's tensile strength;

e, eccentricity of axial load;

A, section area of retrofitted column;

α_1, materials strength reduction coefficient of newly-poured concrete related with stress state of original column; For undamaged original column, $\alpha_1 = 0.9$; for partly damaged or heavily loaded column, $\alpha_1 = 0.7$.

4.4.2 Outer Packing Rolled Angles Method for Retrofitting Brick Columns

The outer packing rolled angles method can obviously improve bearing capacity and resistance to lateral force of brick columns, and the dimensions of columns are not increased too much.

The process of outer packing rolled angles for retrofitting brick columns is as follows. First, the finish coat at four corners of brick should be spudded and cleaned. Then, a 10 mm flat cement mortar should be applied to make level. Meanwhile, rolled angles should be affixed to the surrounding of the loaded brick column and jammed tight by a chucking appliance. After that the rolled angles should be integrated by batten plates. Finally, the chucking appliance should be removed and a cement mortar cover applied to protect rolled angles, as in Fig. 4.10. In order to make sure rolled angles work effectively, they should be well anchored into foundation and the upper part. Furthermore, the size of rolled angles may not be smaller than L50 × 5.

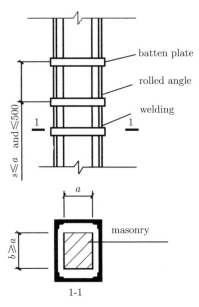

Fig. 4.10   Outer packing angles method for retrofitting brick columns.

A brick column retrofitted by outer packing rolled angles is also a composite brick masonry work whose compressive strength can be increased, for its lateral deformation is constrained by batten plates and rolled angles. In accordance with the calculation methods of concrete composite brick masonry work and reticular reinforcement masonry structure, the bearing capacity can be obtained as follows for the retrofitted column under axial compression:

$$N \leqslant \varphi_{com} \left[fA + \alpha f_a' A_a' \right] + N_{av} \tag{4.11}$$

for the retrofitted column under eccentric compression:

$$N \leqslant fA' + \alpha f_a' A_a' - \sigma_a A_a + N_{av} \tag{4.12}$$

where f_a', design value of compressive strength of retrofitting section steel;

A_a', A_a, section area of compressive and tension retrofitting section steel separately;

N_{av}, the increment of load-bearing capacity of retrofitted brick columns due to the constraint of batten plates and rolled angles, which improves brick masonry's strength, can be calculated by Eq. (4.13):

$$N_{av} = 2\alpha_1 \varphi_n \frac{\rho_{av} f_{ay}}{100} \left(1 - \frac{2e}{y} \right) A \tag{4.13}$$

where ρ_{av}, volume ratio of reinforcement (%). And when the section area of single-limb batten plate is A_{av1}, and the spacing is s:

$$\rho_{av} = \frac{2A_{av1}(a+b)}{abs}$$

f'_a, design value of tensile strength of batten plates;

σ_a, the stress of tension limb of section steel, which can be calculated by Eq. (4.6); and the rest of the symbols are in accordance with the above mentioned.

The height of compression zone x can be calculated by Eq. (4.5).

4.5 Retrofitting of Wall between Windows

The wall between windows is a weak area of masonry structure where cracks or bearing capacity inadequacy usually occurs during an earthquake, vertical loads, uneven settlement and thermal stress. Methods for retrofitting a wall between windows are similar with those for retrofitting the brick wall and brick column, such as setting buttress columns, packing mat reinforcement covers, augmenting section area and outer packing section steels, etc. The detailed retrofitting design methods should be in accordance with the rules outlined in Sections 4.3 and 4.4.

Setting reinforced concrete columns on both sides of original walls could be more effective and economical, and the window apertures can be reduced properly.

When using the method of outer packing rolled angles to retrofit the wall between windows, packing only rolled angles on four corners is not enough. It cannot restrict the middle section of the wall effectively or achieve an expected effect; for the case that the width of the wall is too large compared with its thickness. Therefore, when a wall's ratio of width to thickness is larger than 2.5, a flat iron may be extended vertically on each side of the wall's middle section and tied by the screw bolts, as in Fig. 4.11. After that, a mortar-covering layer should be plastered, which can avoid the rustiness of rolled angles and, meanwhile, achieve the decorative functions of buildings.

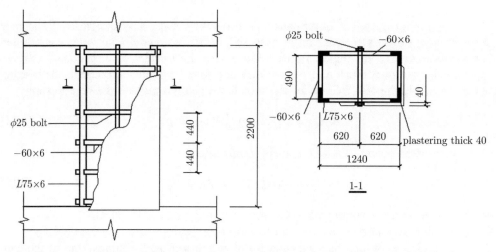

Fig. 4.11   Outer packing rolled angles method for retrofitting wall between windows.

The bearing capacity of a wall between windows retrofitted by rolled angles may be calculated by the same method as that for brick columns retrofitted by outer packing rolled angle.

4.6   Methods for Strengthening the Integrity of Masonry Structures

It is indicated from experimental research and project practice that adding reinforced concrete constructional columns, ring beams and steel pull rods is quite effective for improving building integrity. The methods of adding reinforced concrete constructional columns, ring beams and steel pull rods can be adopted separately, according to the practical situation.

When the original building has no constructional columns, ring beams, or adequate ring beams, it can be retrofitted by adding the constructional columns, ring beams or steel pull rods.

When the original building has inadequate ring beams, it can be retrofitted by adding the ring beams or steel pull rods.

When the original building has adequate ring beams but no constructional columns, it can be retrofitted by adding constructional columns. Meanwhile, retrofitting must also ensure the adequate connection strength and ductility between the constructional columns, and original ring beams or the walls.

When the spacing of the original earthquake-resistant walls is too large, the additional earthquake-resistant walls can be set up first. And then, the retrofitting method of adding constructional columns, ring beams and steel pull rods may be considered.

4.6.1   Detailed Requirements of Additional Constructional Columns, Ring Beams, and Steel Tie Rods

1. Additional constructional columns

(1) Section dimension

In order to ensure the strength of the additional columns and their ability to work with the walls, their section dimensions and reinforcements may not be less than the values listed in Table 4.2 and may be designed in accordance with detailed requirements in Fig. 4.12.

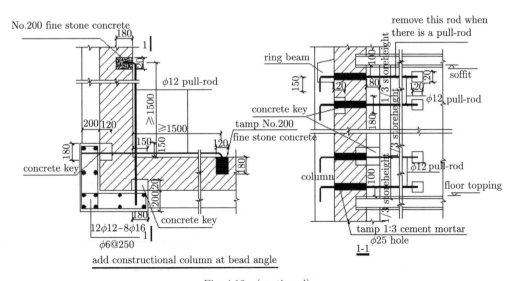

Fig. 4.12   (continued)

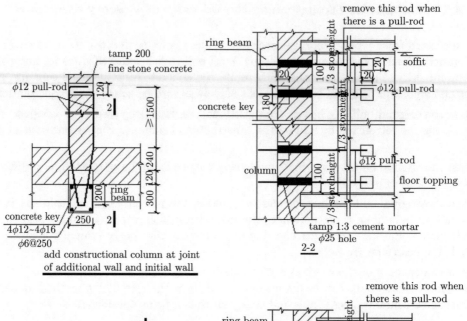

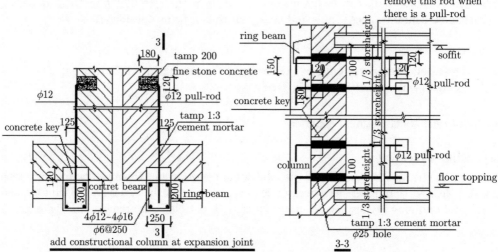

Fig. 4.12 Detailed requirements of additional constructional columns.

The dimensions and reinforcements of the cavity wall's additional columns may be designed in accordance with the requirements listed in Table 4.2.

The connection fit of constructional columns with original beams, ring beams and slabs can be designed in accordance with the details as in Fig. 4.13, Fig. 4.14, and Fig. 4.15, according to the practical situations.

(2) The plinths of additional columns

The plinths must be connected with the originals walls firmly in order to restrict the walls effectively and increase their shear strength and deformability. The embedded depths of the plinths of additional columns may be the same as the exterior wall footing. For example, the embedded depth of the plinth can be 1.5 mm when the exterior wall footing is more than 1.5 mm. The dimensions and detailed requirements of the plinth can be designed in accordance with the details in Fig. 4.16.

2. Additional ring beams

The additional ring beam may adapt to the cast-in-situ reinforced concrete one. Under some special conditions, the section steel ring beam may also be adopted.

Table 4.2 Minimum section dimensions and minimum reinforcements of additional columns

Style of columns	Dimension (Width × height)(mm)	Reinforcement		Cutline
		Main reinforcement	Stirrup	
Rectangular	250 × 150	4 ϕ12	ϕ6 @150~200	
Flat	500 × 70	4 ϕ12	ϕ6 @200	
L-shape	600 × 120, each side	12 ϕ12 double-layers	ϕ6 @200	

Note: Within a distance of 500 mm above and below the floor or roof, the spacing of the additional columns' stirrups should be densified to 100 mm.

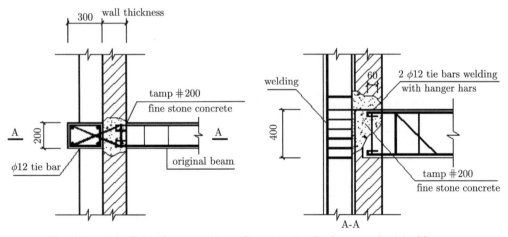

Fig. 4.13 Retrofitting for connections of constructional columns and original beams.

(1) Section dimensions and reinforcements

The section dimensions and reinforcements of additional ring beams may not be less than the requirements listed in Table 4.3.

(2) Additional ring beams should be closed at the same elevation

Ring beams at both sides of the movement joint should be closed. If they meet with an opening, measures should be taken to make them close. And if they meet with brick buttress

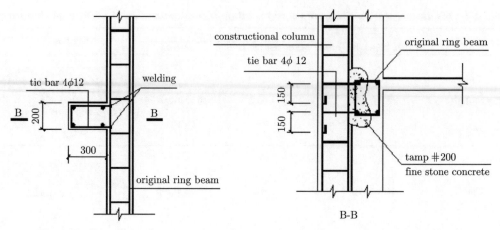

Fig. 4.14   Retrofitting for connections of constructional columns and original ring beams.

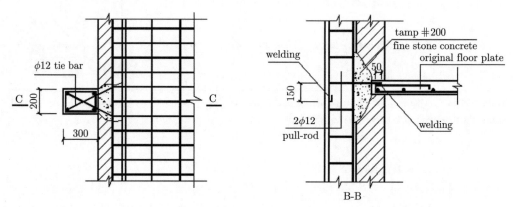

Fig. 4.15   Retrofitting for connections of constructional columns and original slabs.

Table 4.3   Minimum section dimensions and minimum reinforcements of additional ring beams

Ring beam		Earthquake intensity			Remarks
		7	8	9	
Reinforced concrete	Dimension (Height × width) (mm)	180 × 120			Height is dimension parallel to wall surface. Width is dimension vertical to wall surface
	Main reinforcement	4ϕ8	4ϕ10	4ϕ12	
	Stirrup	ϕ6@200			When ring beam is connected with additional column, spacing of stirrups may be densified to 100 mm, within 500 mm of each side of column
Section steel	Channel steel rolled angle	8 [75 × 6]			

columns and down pipes, measures should be adopted to keep the continuity of ring beams.

(3) Ring beams should be connected effectively with masonry walls

In order to ensure a reliable bonding of masonry walls and the concrete ring beams, the surfaces of the walls at connection sites should be cleaned before ring beams are laid. Meanwhile, common anchor bolts, reinforced concrete pin keys and other connecting pieces can also be used.

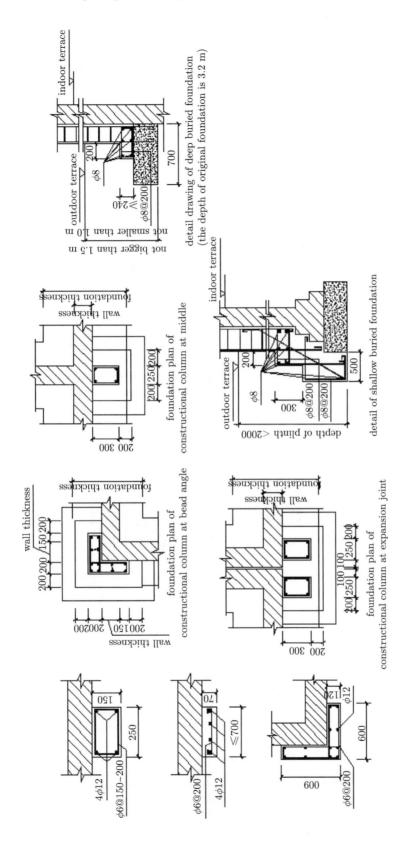

Fig. 4.16　Retrofitting for pinth of additional constructional columns.

When the additional reinforced concrete ring beam is connected with the wall by common anchor bolts, one end of the anchor bolt should be embedded into the ring beam in the shape of a right-angle hook, and the embedded length should be $30\,d$ (d is the diameter of the anchor bolt). And the other end should be fastened by the nut, as in Fig. 4.17.

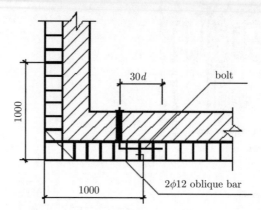

Fig. 4.17   Retrofitting for connections of ring beam and wall with common anchor bolts.

When the additional reinforced concrete ring beam is connected by pin keys, the height of pin keys should be the same as ring beams, and their width should be 180 mm. The insert depth of pin keys in walls should not be less than 180 mm or the thickness of walls, and the reinforcement should not be less than $4\phi8$, the spacing may be 1000~2000 mm. Pin keys of exterior walls may be set on both sides of windows, and it is necessary to avoid the damage of walls when chiseling holes for pin keys, as in Fig. 4.18.

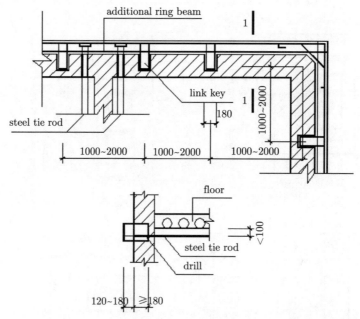

Fig. 4.18   Retrofitting for connections of reinforced concrete ring beam and pin keys.

Section steel ring beams should be connected with walls by common bolts at intervals of 1000~1500 mm and the diameter of bolts should not be smaller than $\phi12$. And the gap between ring beams and walls can be plugged firmly by dry cement mortar.

3. Steel tie rods

Steel tie rods may not be smaller than $2\phi14$, and they should also be connected effectively with additional ring beams. The ends of steel tie rods may be welded with steel underboardings, and then be embedded into ring beams as in Fig. 4.19. The connection can also be achieved by other methods.

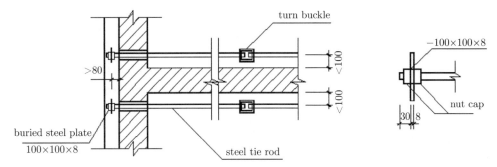

Fig. 4.19   Retrofitting for steel tie rods.

In order to tense steel tie rods, a turn buckle is usually set on the middle of rods. And rust should be removed from steel tie rods, then one cover of anti-rusting paint and two covers of aluminum paint should be applied.

4. Connection requirements of columns and walls

Additional columns may be connected to cross walls with steel tie rods, pin keys or other connecting pieces at the 1/3 and 2/3 story heights of each floor. And plinths should also be connected with wall footings by pin keys or other connecting pieces at outdoor terraces and the contact points of walls and pedestal footings.

5. Other requirements

When there is no additional column at the walkway of a middle corridor building, in order to ensure a fine work state between the additional columns and the walls on both sides of the walkway, cast-in-situ reinforced concrete beams or composite steel beams may be added at the axes of the cross walls with additional columns under the floor (roof). Meanwhile, the section dimensions of beams should not be smaller than 240 mm × 300 mm and the insert depth of both ends into the cross walls may not be less than 600 mm. Besides, the continuity of longitudinal bars in the additional columns should be ensured. When concrete strength grade is more than C20, grade I reinforcements may be adopted. And in the positions where there is no cross wall, additional columns should be bonded firmly with the depth beams in the depth direction of floors (roof) or with the cast-in-situ floors (roof).

4.6.2   Calculations of the Shear Strength of Walls Retrofitted by Additional Columns, Ring Beams, and Steel Tie Rods

It is indicated from research that the increase of the shear strength of walls retrofitted by additional constructional columns is limited and is usually no more than 30%. But it can increase overall anti-collapse capability by 50%~100%.

When all the detailed requirements of additional constructional columns, ring beams, and steel tie rods outlined in this chapter are complied with, the shear strength of walls retrofitted by additional reinforced concrete columns can be calculated in line with the following Eq. (4.14):

$$V_k \leqslant \zeta_N f_v A_{mk} + [(1 + \alpha_s) f_t A_c + 0.4 f_y A_{st}] \eta_g \qquad (4.14)$$

where V_k, seismic shear force borne by k retrofitted wall;

f_v, design value of masonry walls' shear strength;

α_s, influence coefficient of the additional columns' longitudinal reinforcements,

$$\alpha_s = 0.4\frac{A_s}{A_c} \cdot \frac{f_y}{f_t}$$

f_t, design value of concrete tensile strength of additional columns;

A_c, section area of additional columns;

A_{st}, section area of the extended steel tie rod added on the additional columns of k wall;

f_y, design value of the tensile strength of the steel tie rod added on the additional columns of k wall;

ζ_N, normal stress influence coefficient of masonry strength, which should follow *Code for Seismic Design of Buildings*;

A_{mk}, effective cross-section area of brick walls;

η_g, work participating coefficient of constructional columns.

For the wall with constructional columns at both ends and ring beams on its top and bottom, if its height/width ratio is no less than 0.5, η_g should be 1.1; if the ratio is no more than 0.5, η_g should be 1.0.

For the wall with constructional column just at one end and ring beams on its top and bottom, η_g may be 0.9; for the wall with constructional columns on both ends and a door or a large opening on it, it can be considered as that with constructional column only on one end, and η_g should be 0.9.

For the wall with columns on both ends, a ring beam only on its top (or bottom) and steel tie rod on its bottom (or top), η_g may be 0.9. If there is no steel tie rod, η_g should be 0; And if there is no ring beam on either its top or bottom, the contribution to wall's shear strength from constructional columns can be ignored, and η_g should be 0.

If ground floor has no foundation ring beam and foundation is connected firmly with constructional columns, because of the foundation's firm constraint, the ground floor walls can be considered as those with foundation ring beams.

The contribution to shear strength from additional constructional columns, ring beams and steel tie rods, which can be calculated by Eq. (4.14), may not be more than 30% of that of the original wall.

4.7   Retrofitting Methods for Connections between Masonry Members

When connection strength of masonry members is inadequate, it will lead to partial crack, which may cause partial collapse under earthquake action or uneven settlement.

(1) Retrofitting for connections between additional walls and original floors or walls

The retrofitting method can be designed, as in Fig. 4.20 and Fig. 4.21.

(2) Retrofitting for connections between exterior and inner walls

When connections between exterior and inner walls cannot comply with the requirements, they can be retrofitted by binding reinforcements, as shown in Fig. 4.22.

(3) Retrofitting for connections between exterior walls and floors (roof)

When connections between exterior walls and floors (roof) cannot comply with the requirements, they can be retrofitted, as in Fig. 4.23.

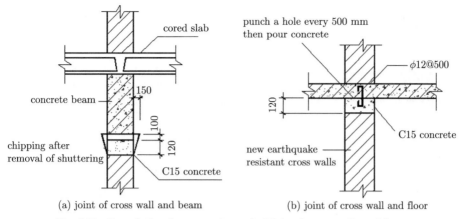

(a) joint of cross wall and beam   (b) joint of cross wall and floor

Fig. 4.20   Retrofitting for connections of additional cross walls and floors.

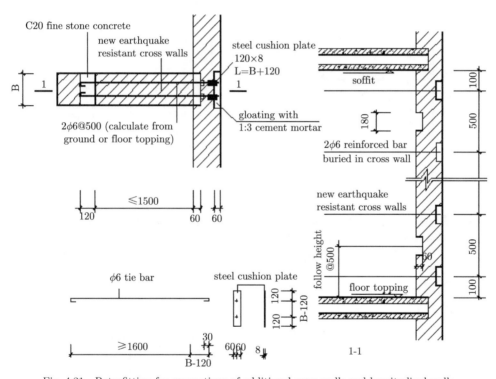

Fig. 4.21   Retrofitting for connections of additional cross walls and longitudinal walls.

(4) Retrofitting for connections between exterior walls and ring beams

When connections between exterior walls and ring beams cannot comply with the requirements, they can be retrofitted, as in Fig. 4.24.

(5) Retrofitting for connections between exterior walls

When connections between exterior walls can not comply with the requirements, reinforced concrete or mat reinforced-cement mortar covers can retrofit the inside corners of exterior walls, as in Fig. 4.25. Also, steel plates can be adopted to retrofit the corners of exterior walls, as in Fig. 4.26.

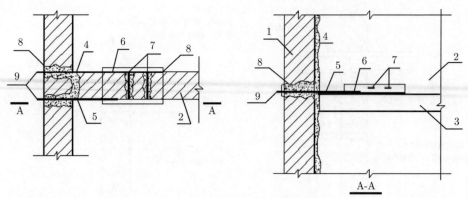

Fig. 4.22   Retrofitting for connections of exterior walls and inner walls by binding bars:
1. exterior wall; 2. inner wall; 3. slab; 4. cracks at connection site (filled with mortar); 5. tie rod welded
on rolled angles; 6. rolled angle; 7. screw bolt; 8. porthole (filled with mortar after putting in the tie rod);
9. tightened by screw nut.

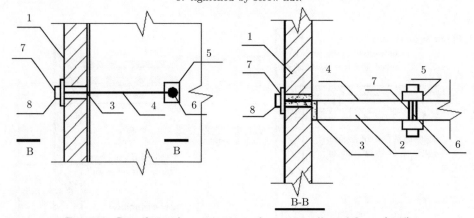

Fig. 4.23   Retrofitting for connections of exterior walls and floors (roof):
1. exterior wall; 2. reinforced concrete slab; 3. cracks between wall and slab (filled with mortar);
4. tie rod welded on steel plate; 5. steel plate; 6. screw bolt; 7. portholes in the wall and the floor (filled
with mortar after putting the tie rod and screw bolt); 8. tightened by screw nut.

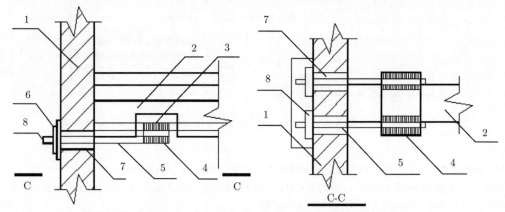

Fig. 4.24   Retrofitting for connections of exterior walls and ring beams:
1. exterior wall; 2. reinforced concrete slab; 3. bare reinforcement; 4. steel plate welded on bare
reinforcement; 5. tie rod welded on steel plate; 6. underboarding to fix the tie rod; 7. portholes for
inserting tie rod (filled with mortar after putting the tie rod and screw bolt);
8. tightened by screw nut.

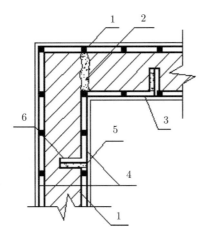

Fig. 4.25   Retrofitting inside corner of exterior walls by mat reinforcement mortar covers:
1. corner of exterior wall; 2. cracks at connection site (filled with mortar); 3. holder of mat reinforced-cement mortar covers and reinforcement concrete; 4. mat reinforcement; 5. anchor rod made by deformed bar, with a diameter of 10 mm and a space of 600~800 mm at vertical and horizontal directions; 6. porthole in the wall, with a depth of no less than 100 mm.

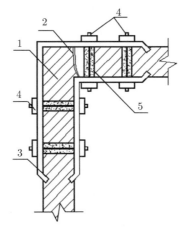

Fig. 4.26   Retrofitting inside corner of exterior walls by steel plates:
1. corner of exterior wall; 2. cracks at connection site (filled with mortar); 3. dual steel plates made by steel rods; 4. tightened by screw nut; 5. porthole in the wall (filled with mortar after installing screw bolt).

CHAPTER 5

Retrofitting Design of Wood Structures

5.1 Introduction

Wood structures are mainly small and medium size buildings. Common types of load-bearing wood structures include wood frame with dougong, old-type wood frame, wood structure with roof truss on columns, wood purlin structure, and Tibetan wood structure.

As an organic material, the natural flaws of timber (knag, crack, warp, etc.) may deteriorate during the service life. In addition, various degrees of damage can be caused by some extraneous factors such as deficiency of design and construction, fungus attack, pests, chemical corrosion, improper management and natural disasters. Strengthening is accordingly needed on timber beams, roof trusses, columns and other wooden members.

When existing wood buildings are rated insufficient for seismic resistance, they must be strengthened according to seismic requirements.

5.1.1 Reasons for Retrofitting of Wood Structures

(1) Hazards caused by wood defects

Knag, twill, crack and warp, which are all wood defects, can weaken timber strength in various degrees according to their different size and position, while some defects can even endanger the bearing capacity of wood structures. It is necessary to conduct retrofitting. Fig. 5.1 shows some examples of hazardous defects.

(2) Pest and fungus attack

Pests are hazardous to wood structures, and include various kinds of insects, especially termites. Fungus attack is a kind of corrosion hazard caused by wood-decay fungi. And the mechanical properties of timber will be changed after corrosion, which will lead to the damage of wood structures.

(3) Chemical corrosion

In modern industrial production, some factories (e.g. acid pickling workshops, dye bleaching workshops of textile factories and chemical workshops with corrosive gases, etc.) will generate corrosive gases which can erode wood structures, weaken their strength and make them unsatisfactory and dangerous for regular service.

(4) Wind and earthquake disaster

Under extremely severe wind action, wooden roofs may be blown off, wooden columns may be broken and any original defect may also be exacerbated.

Under earthquake action, damage to wood buildings may vary according to different local timber properties, member sizes, constructional methods, and construction qualities. The most common damages are:

a. Loosening, pulling-out or splitting of tenons, and separation of partial members.

b. Inclination of timber frame, shift of plinths, and collapse of enclosure walls.

c. Instability of structure caused by failure of structure bracings.

d. Worsening of original damages due to earthquake action.

(5) Changes of service requirements of the wood structures

Structures should be strengthened before their service requirements are changed or they are rebuilt, which may lead to an increase of loads or irrational stress concentration.

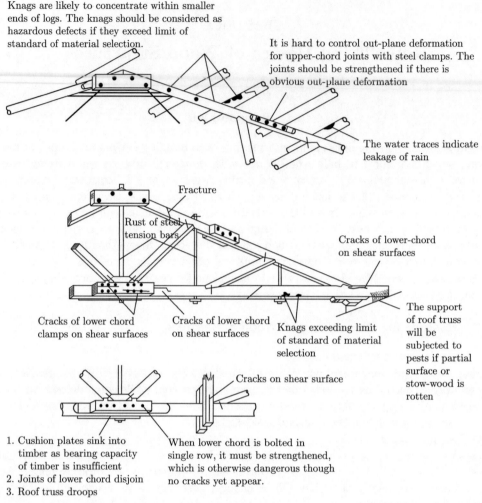

Knags are likely to concentrate within smaller ends of logs. The knags should be considered as hazardous defects if they exceed limit of standard of material selection.

It is hard to control out-plane deformation for upper-chord joints with steel clamps. The joints should be strengthened if there is obvious out-plane deformation

The water traces indicate leakage of rain

Fracture

Rust of steel tension bars

Cracks of lower-chord on shear surfaces

Cracks of lower chord clamps on shear surfaces

Cracks of lower chord on shear surfaces

Knags exceeding limit of standard of material selection

The support of roof truss will be subjected to pests if partial surface or stow-wood is rotten

Cracks on shear surface

1. Cushion plates sink into timber as bearing capacity of timber is insufficient
2. Joints of lower chord disjoin
3. Roof truss droops

When lower chord is bolted in single row, it must be strengthened, which is otherwise dangerous though no cracks yet appear.

Fig. 5.1   Examples of hazardous defects.

(6) Deficiency of design and construction

Excessive stress and careless construction details are common faults in structural design.

During construction, the most common quality defects include loose joints, deficient member strength caused by dimension errors, rickety roof-truss joints, split members and insufficient constructional requirements for mortises, etc.

5.1.2   Principles of Wood Structure Retrofitting

The methods of wood structure construction vary by locations and ages. Accordingly different strengthening methods are adopted in different districts or projects. There is no certain set of methods to follow. On the premise of meeting service demands, the strengthening method should consider local conditions and eliminate hazards economically so as to make the wood structures safe for service. The main points of retrofitting are as follows:

(1) Reliability assessment

Before strengthening, reliability of the original structure and members should be assessed, and service conditions of structure and material should be investigated. The contents include:

a. Mechanical and material properties of lumbers used in the wood structure.

b. Degree and characteristics of the influence on strength caused by defects of wooden components.

c. Inspection of the positions and characteristics of corrosion and pests, and relevant analysis of their damages to the structure.

d. Temperature and humidity conditions where the wood structure is located.

e. Understanding and measurement of the chemical component containing corrosive medium in the surroundings in which the wood structure is located.

f. Stress and deformation state of bearing members and functional modes of connections at main joints.

(2) Elimination of hidden hazards

Before retrofitting, reasons for wood structure retrofitting must be analyzed. Crucial issues that endanger the safety of the wood structure should be solved firstly, followed by relevant solutions proposed to deal with other damages. Sufficient consideration should be given to the secondary damages resulting from the same causes so that the hidden hazards can be thoroughly eliminated.

(3) Strengthening materials

The lumber and steel selected for wood structure retrofitting should conform to the relevant current national standards. The connecting lumber, used for strengthening bearing members, should be in straight grain without defects, and its water content should be strictly controlled.

Utilization of old lumber for strengthening bearing members or reuse of old members (wooden columns, purlins, joists, etc.) must be verified to conform to the relevant standards and design requirements.

(4) Consolidating and improving the structure load resisting system

Depending on different structure loading states, retrofitting requirements, material supplies, construction sites and available construction conditions, the retrofitting of wood structures has two levels: consolidating and improving the current situation. Here the current situation mainly refers to the conditions of structure load resisting systems.

a. Consolidating the current situation

The structure is in a good condition as a whole; however, hidden hazards may endanger the safety and serviceability of the structure if no attention is given to them. Efficient measures should be adopted to eliminate local hazards or control the worsening of hazards so as to ensure safety and maintain reliability of original structure.

This level of retrofitting focuses on the wood structure itself. For example:

a) Failure of the individual members in roof trusses or braces, or damage of the clamps at joints, can be solved by replacing with new members or plates which conform to the standards.

b) When damages take place in some parts of members and joints, or defects in some members are over the limit, local strengthening can be efficient to meet the safety demand.

c) If the lumber near the bearing area is eroded only on the surface and in an early stage, with the inner layers in good condition and dry, it is feasible to remove the rotten parts thoroughly and apply antiseptic to the surface and inner layers. Then the structure can maintain normal service.

d) The original wooden tension rods can be replaced by steel ones, if they are disabled.

e) Skew of the roof truss or large out-of-plane deformation of individual compression rods can be rectified by angle steels and bolts to meet the service demand. Braces may be added when necessary.

b. Improving the current situation

When there is insufficient spatial stiffness or bearing capacity of structure or members due to overload, change of service condition, design or construction errors, structural member defects, etc., it is necessary to conduct retrofitting according to the main hazards, or even rebuild the whole structure.

This level of retrofitting focuses not only on wood structure itself, but also on its loads, bearing states and even the surroundings as well. For example:

a) If the timber surfaces at roof trusses or column bearings are eroded by humidity and enclosed, the rotten parts should be removed and antiseptic chemicals should be applied. It is more important to modify the configuration at the bearings to eliminate moisture.

If the bearing part has been eroded severely, measures should be taken to replace the rotten members with new ones. If moisture proofing of the supporting parts is hard to attain, the rotten parts should be replaced with rolled steel components or precast reinforced concrete joints.

b) Wooden roofs are vulnerable to rot in an environment of high temperature, humidity and poor ventilation. This may cause corrosion and additional deflection of the wood structure. Besides retrofitting the structure, it is preferable to improve the surrounding condition and keep the structure in a dry and normal environment.

c) If the members are overburdened because of excessive load or service alteration, the original members should be replaced by new ones with much lighter materials. One or two columns can be added to the structure, if possible, to shorten the span and reduce the stress of members. Then more attention should be paid to the changing of stress in original structural members caused by the added columns.

d) Rebuild, strengthen or extend spatial bracing system that is deficient in the original structure.

e) Strengthen and modify the original bearing components such as walls and columns to ensure the normal serviceability of wood structure.

(5) Main points of retrofitting designs of wood structures

a. Economic effects should be taken into comprehensive consideration when conducting retrofitting designs. The original structure should be damaged as slightly as possible, while the serviceable structural members should be reserved.

b. Structural calculating sketches should be in accordance with actual load and stress on the structure.

c. After long-term service, the material strength of the lumber will decrease to some extent. Therefore the material strength should be able to meet the real condition.

d. The actual state of structural deformation, cross section changes, increase or decrease of members and nodal displacements should be accurately determined as the basis of the calculation.

e. The favorable conditions of original structure should be used to improve the irrational structures.

f. Steel is preferable to timber. It is better to make tension rods and clamps with steel.

(6) Main points of retrofitting constructions of wood structures

Construction schemes should be formulated before retrofitting constructions. Retrofitting should be executed in the sequence of supporting first and strengthening second.

Make full-scale mould boards according to actual dimensions of designs and members, and number them. Manufacture the strengthening members according to the mould boards.

When wooden clamps are used, the materials for retrofitting and the diameters, quantity, and position of bolts should conform to design demands. When the holes for splicing are drilled, related components should be positioned, fastened temporarily and drilled through at one time, to ensure that the positions of holes on each member are consistent. The diameters of holes should not be larger than those of the shear bolts by 1 mm, or larger than the fastening bolts by 2 mm.

The splices of the round steel tension bars for retrofitting should be welded by double ties. The diameter of the round steel tie ought to be no less than 0.75 times that of the tension bar. The tie length on one side of splice should be no shorter than 5 times the diameter of tension bar.

(7) Seismic strengthening of wood structures

a. Seismic checking calculations are not necessary in seismic strengthening of wood structures.

b. Seismic capacities of wood frames should be improved by strengthening. Based on actual condition, some efficient measures can be adopted such as reducing roof weight, strengthening wood frame, reinforcing joints, adding column bracings and brick walls and so on. The column bracings and seismic walls should be set uniformly in floor plane. Bearing systems are the key points of seismic strengthening.

5.2   Retrofitting of Wooden Beams

1. Strengthening with clamped or haunched connection methods

The supporting points (the ends connected with walls) of wooden beams are vulnerable to decay, corrosion and other damages. It is reliable to adopt clamped or haunched connection methods to strengthen the wooden beams in these cases. If the damage depth on the upper or bottom sides of the beam is larger than 1/3 of its height, it should be strengthened with clamps. Otherwise, if damage depth is larger than 3/5 of the height, the beam-end should be replaced by a new one. If the beam-end is moth-eaten, it can also be strengthened with clamps according to calculation.

Before strengthening, the beams should be supported temporarily or unloaded. The supporting points should be in alignment when the beams are within different stories. The damaged parts of the supported beams should be cut off before strengthening with the methods mentioned above.

(1) Clamped connection

Beams can be strengthened with clamped connections (Fig. 5.2). The cross section and

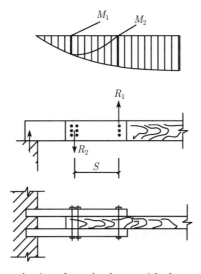

Fig. 5.2   Strengthening of wooden beam with clamped connection.

material property of the wooden clamps should exceed those of the original beam. The clamps should be made of straight grained and air-dried timber without knag or boxed heart. Under no circumstance can humid timber be used. The damaged end of the beam should be cut even and be connected with the substitute timber tightly and straightly. The interface between clamps and beam should be smooth and close when fastened by bolts. For a round-section beam, the clamps and the newly processed plane should be connected well.

Specification and quantity of bolts and length of clamps should be calculated according to current codes.

The force on bolts can be calculated by the following formula:

$$R_1 = \frac{M_1}{S}, \quad R_2 = \frac{M_2}{S} \tag{5.1}$$

where S is the distance between force R_1 and R_2 on the bolt; M_1 is the moment at the R_2 point (the moment within the wooden clamps); M_2 is the moment at the R_2 point (the moment within the beam).

(2) Haunched connection

Beams can be strengthened with channel steel or other material support at the bottom (Fig. 5.3). The tension bolts connect channel steel and timber beam, and their cushion plate should be calculated for checking. Tensile force within the bolt can be calculated by the following formula:

$$R_1 = \frac{M_1}{S}, \quad R_2 = \frac{M_2}{S} \tag{5.2}$$

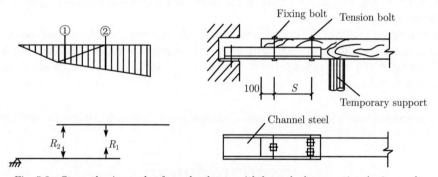

Fig. 5.3   Strengthening ends of wooden beam with haunched connection (unit: mm).

where S is the distance between reaction force R_1 and R_2; M_1 is the moment in the channel steel (equal to the moment at beam section ①); M_2 is the moment at beam section ②; R_1 is the force within the tension bolt; R_2 is the extrusion force between channel steel and end across-grain plane where the bolt is fixed.

It is reliable to use channel steel as a haunched connection because its configuration is easy to handle. This method can be adopted when a beam is difficult to strengthen by wooden clamps.

2. Strengthening with bottom-bracing steel tension rods

There are diverse strengthening forms with bottom-bracing steel tension rods. One simple form is shown in Fig. 5.4. It can be applied to strengthen a shaky beam with small cross section, which has deficient bearing capacity or excessive deflection. The beam and the bottom-bracing steel tension rod constitute a new bearing member.

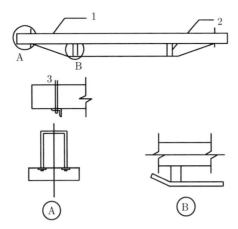

Fig. 5.4   Strengthening of beam with bottom-bracing steel tension rods:

1. wooden beam; 2. brace rod; 3. steel tension rod.

Before strengthening, it is necessary to check whether the ends of the beam are corroded or moth-eaten. The steel tension rod can only be fixed well if it is of good material quality.

Make samples of the steel items, tension rods, and brace rod according to design demands and actual dimensions of strengthening members. Checking must be conducted before application. Temporarily support and fix each component during application. Do not fix the bracing rods until the trial assembly reaches the design requirements. Then pull the tension rods tight. The steel rods should be pulled tightly and straight, and be fastened firmly and the interfaces of beams and bracing rods or steel items should be inosculated. New bottom bracing should be added within the same vertical plane to the beam axis.

3. Strengthening with flat steel hoops

Strengthening with flat steel hoop is suitable for wooden beams subjected to longitudinal splitting damage (Fig. 5.5).

The flat steel hoop should be in accurate dimension and uniform configuration, and adhered to the beam. The lofting should be in full size, while the bolts need to be fastened and fixed individually with no loosing of the steel hoops. It must be noted that there should be a gap after the bayonet bolt closes. Only in this way can the bolt be fastened tightly and adhered to the beam well. The cracks on the beam should be stuffed.

4. Strengthening with double clamps

When double clamps are used for strengthening, the lofting and fabrication should be in accordance with design requirements and real dimensions. The clamping boards should be parallel and symmetrical, with correct angles and positions of bolts. The interface between both ends of clamping boards and beams (or columns) should be inosculated.

5. Strengthening beam-column joints with bolster

In this case (shown in Fig. 5.6), tenons at joints should be reset first, and then wooden wedges driven in to achieve fixation. The holes on the bolsters and columns should be drilled through at one time. Fastened to columns by the bolts, the bolsters should be attached to the beams and columns closely.

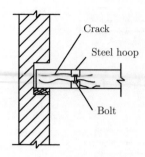

Fig. 5.5   Strengthening with flat
steel hoops.

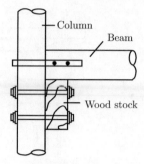

Fig. 5.6   Strengthening beam-column
joints with bolsters.

5.3   Retrofitting of Wooden Roof Trusses

When wooden roof trusses need strengthening or wood frames need rectification, it is mainly because the timber is subjected to dampness or pest infestation, which results in partial failure or severe deformation. Since the wooden roof trusses and wood frames are both important bearing components, it is necessary to establish different construction schemes and accident prevention measures to ensure safe construction before strengthening.

5.3.1   Strengthening with Wooden Clamp

When fractures appear at upper chords of roof trusses due to over large cross grain or harmful knags appear at partial joints, it is feasible to strengthen them with wooden clamps (shown in Fig. 5.7). It is less favorable when these problems emerge at the internodes of upper chords. In this case, it is possible to attach a new piece of timber under the defective upper chord. It should be noted that, bearings of both ends of the new timber should be reliably treated to help the upper chord bear load effectively.

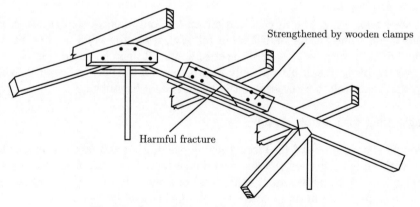

Fig. 5.7   Strengthening of fractured upper chord.

The upper chords and diagonal web members are strengthened by clamps, as shown in Fig. 5.8.

When the lower chords of roof trusses have severe corrosion and long damaged segments, it is feasible to cut all corroded segments down, replace them with new timber, and then bolt wooden clamps or steel clamps (Fig. 5.9).

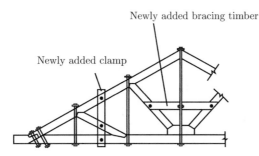

Fig. 5.8   Strengthening of upper chord and diagonal web members with clamps.

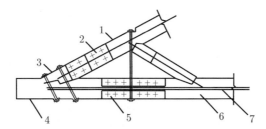

Fig. 5.9   Strengthening of rotten joints:
1. original upper chord; 2. newly added clamps; 3. new upper chord segment; 4. new lower chord segment;
5. newly added steel clamps; 6. original lower chord; 7. axis of refined gross cross section.

5.3.2   Strengthening with Clamps and String Rods

When strengthening with clamps and rods, the construction should be conducted in a sequence of fixing clamps, applying additives, fixing steel components, and stretching string rods. Steel components and wooden components should be combined tightly at the bearing surface and located accurately, while string rods should be straight, reliable, parallel and symmetric. For bolts for round steel string rods, twin nuts must be adopted, while the extension bolt length of the nut should be no less than 0.8 times its diameter. Requirements of strengthening roof trusses with clamps and string rods are listed as follows:

a. When cracks (no matter what the width is) appear on the shear surface of tensile joints of lower chords, or the lower chord is bolted just by a single row and other regions of lower chord are intact, it is feasible to adopt new tension instruments to replace the original bolt connections. If clamps and string rods are used to strengthen roof trusses (Fig. 5.10 and Fig. 5.11), favorable direction and position of clamp must be selected, while the weakening impact of clamp bolt on lower chords should be calculated for checking.

b. When the roof truss ends are severely decayed and the original timber at the joint cannot be utilized, strengthening can be conducted as the following methods:

① If the condition resulting in decay can be rooted up, it is feasible to cut off the decayed region and replace it with new timber, as shown in Fig. 5.12.

② If the condition resulting in decay cannot be eliminated, it is feasible to cut off the decayed region and replace it with rolled steel components (Fig. 5.13), or to replace the wooden joint with a reinforced concrete joint. If there is a parapet, the reinforced concrete ends for strengthening may be taken into consideration with the gutter (Fig. 5.14).

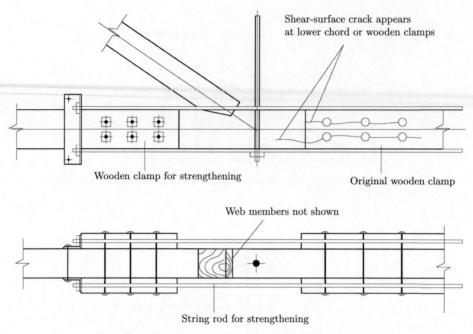

Fig. 5.10   Strengthening of tensile joints at lower chord I.

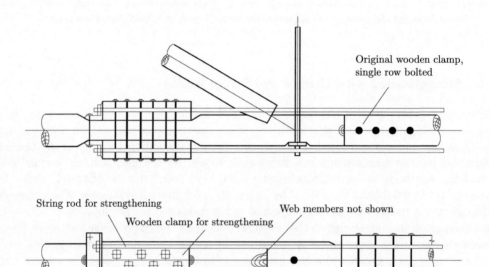

Fig. 5.11   Strengthening of tensile joints at lower chord II.

5.3.3   Strengthening Vertical Wooden Rods with Steel Tendons

When cracks appear on the shear surfaces of the vertical wooden tension elements of roof trusses, it is feasible to replace the vertical wooden rods with a new round steel tension rods, installed as close to the original wooden tension element as possible. Wooden tension elements can be strengthened by steel tension rods not only in the center of roof trusses but also at the internodes (Fig. 5.15).

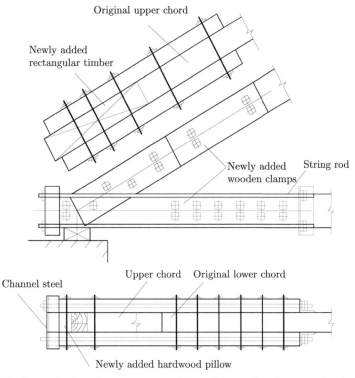

Fig. 5.12   Strengthening of rotten support joints with wooden clamps and string rods.

The steel tension rods generally consist of two or four members. The string rods in this form must be driven through the eyelets of steel components, and parallel to the original vertical wooden rods. The steel components should be machined regularly and fastened tightly to the roof truss. And string rods should be straight and firmly fixed. Before the middle vertical wooden rod is strengthened by steel tension rods, part of the roof should be removed and the ridge purlin should be supported temporarily. Vertical wooden rods at internode can be strengthened by steel tension rod without removing roof.

5.3.4   Rectification of Wood Frames

Wood frames are monolithic load bearing structures. Due to the corrosion at the bottom of columns and long service without repair, the building may incline at one or two sides. Since this problem severely affects regular service and structural safety, it is feasible to conduct rectification. Constructional sequence is the key for rectification: loosening, synchronizing, intermitting, and resetting.

Before rectification, floors and roofs should be unloaded, and then partial masonry connected to the wood frame should be separated. Then towing points should be disposed appropriately on the wood frame, and be connected by hauling ropes. Generally, the supporting timber at the bottom of columns should be reliable and fixed, and the hauling ropes, pull-back ropes and stretching instruments must have sufficient strength. If the original frame has defects, strengthening should be conducted first. Before the application, an observation instrument must be set up. The rectification cannot be conducted unless the outcome of trial stretching conforms to the requirements.

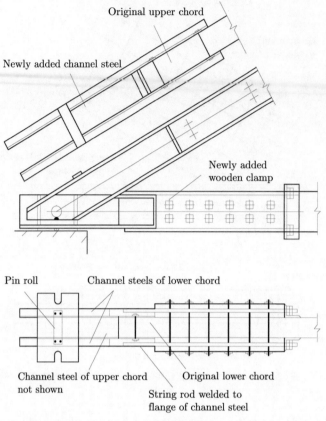

Fig. 5.13   Strengthening of rotten support joints with rolled steel components.

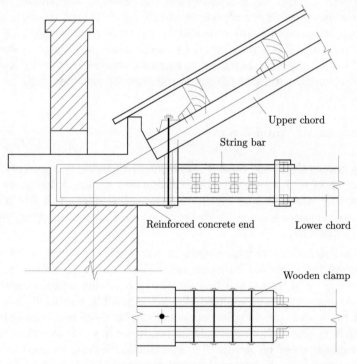

Fig. 5.14   Strengthening of rotten support joints and adding cornice with reinforced concrete.

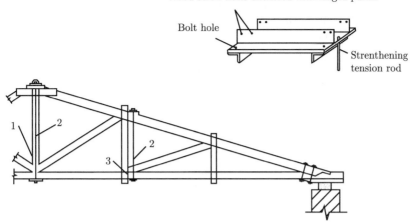

Fig. 5.15   Strengthening of vertical wooden rods:
1. original wooden rods; 2. strengthening tension rods; 3. cracks on the shear surface of original clamps.

When rectification is carried out, the hauling ropes and pull-back ropes must be run synchronously with intermissions. The rectification can only be continued when rectified quantity and structural state are normal. During the application, perpendicularity of the frame and variation of the joints must be observed and recorded. Generally, the overcorrection should not exceed 20 mm, and the rectified frame should be vertical and stable for acceptance.

After rectification, some work such as repairing and fixing connective joints, masoning walls, and repairing roofs should be done. Then the rectification tools can be removed, with every column axis of the wood frame vertical and in the same vertical plane.

When the rectification is for two-story frame buildings, additional towing points, pull-back ropes and stretching instruments should be installed in accordance with the building situation. If the frame has double-direction inclination, one direction should be rectified to conform to design requirements before the other direction is rectified.

5.4   Retrofitting of Wooden Columns

When buildings show settlement due to deformation or corrosion and damage at the bottom of columns, the wooden columns must be propped vertically before strengthening. The wooden columns embedded in walls should generally be strengthened after masonry removal. Before strengthening, beams and trusses upon the columns should be supported temporarily.

It is feasible to adopt methods such as jointing columns or installing plinths to strengthen the wooden columns corroded at the bottom.

Use bricks or concrete to build plinths. The cut-down sections of columns should be perpendicular to the axial lines of the columns. At the joints of the columns and the plinths, prevention against corrosion and humidity should be carried out. Temporary supports cannot be removed unless the concrete strength of the plinth is above 50% of the design strength. The joint surface between the column and the plinth should be smooth and coupled tightly. Specification, dimension, disposition and embedded depth of the anchoring steel components should conform to the design requirements. Eyelets, connecting steel components and wooden columns should be drilled through with bolts tied up.

When columns are jointed with timber, it is feasible to adopt flat-bottom connection or splicing-tenon methods.

When flat-bottom butt connection is adopted, the compressive surface cut down should be perpendicular to the axis of the column, smooth, and tightly coupled. And clamps should be connected to the column and fixed reliably.

When splicing-tenon (Fig. 5.16) is adopted, both upper and lower compressive surfaces should be tightly coupled after being fixed with bolts. Vertical interface should be at the location of the axis of column.

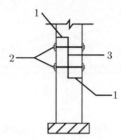

Fig. 5.16   Splicing-tenon method:
1. compressive surface; 2. bolts; 3. vertical interface.

5.5   Retrofitting of Other Wooden Components

Generally speaking, temporary supports must be installed before wooden purlins, stairs, and ceilings are strengthened.

5.5.1   Strengthening of Wooden Purlins

(1) Strengthening of attached purlins

Selecting of purlins should consider design and condition of buildings. When attached purlins are supported by brick walls, holes chiseled on brick walls should be uniform and close to the original damaged purlins, while the ends of attached purlins embedded in the brick walls should be treated to prevent corrosion and be tightened by wooden wedges. The attached purlins should be adhered to the substrate of the roof above and tightened by wooden wedges if the adherence is not steady; holes on the walls should be filled and the support lengths of purlins should conform to design requirements.

When attached purlins are supported by roof trusses, the depth of groove should not be larger than 1/3 of the height of the purlin if its ends are grooved; if the attached purlins are supported by bolsters, the connections between the bolsters and the upper chords of roof trusses should be fixed reliably, and the support length should conform to design requirements.

In areas that are prone to earthquakes, typhoons or strong winds, attached purlins should be anchored to roof trusses or walls reliably to conform to relevant specifications.

(2) Replacing purlins with bracketing beams

When purlins deform severely due to insufficient section areas, it is feasible to adopt this strengthening method. A round or rectangle timber is installed under the purlin. As a result, the timber and the strengthened purlin compose a kingpost truss, as the purlin becomes one member of the truss (Fig. 5.17).

(3) Strengthening with round tension rods

First, drill one $\phi17$ hole transversely 10 cm away from each end of the purlin that needs strengthening. Then, install the precast units, including two $\phi8$ steel loops, one $\phi12$ rebar, two $\phi16$ bolts, two 50 mm $\times$ 50 mm $\times$ 6 mm steel angles and nuts, as shown in Fig. 5.18.

Finally, screw down the nuts. If the purlin deflects overly, lift it a little before screwing the nuts. This method is simple and convenient in application.

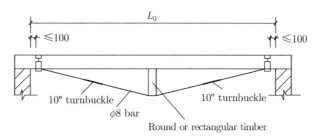

Fig. 5.17   Strengthening of purlin
(replacing purlin with bracketing beam, unit: mm).

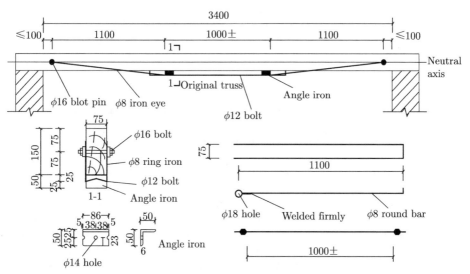

Fig. 5.18   Strengthening of purlin (with round tension rods, unit: mm).

5.5.2   Strengthening of Wooden Stairs

Most damages of wooden stairs are caused when the close-to-ground part of the skew beam is subjected to humidity or insect pests, which results in settlement of stairs. Additionally, there are other damages such as loose trigonal timber resulting from long service.

Wooden stairs to be strengthened are of two types, namely exposed beam and hidden beam stairs. For exposed beam stairs, it is feasible to replace the trigonal timber or add splints. As for hidden beam stairs, the damaged skew beams are usually replaced by new ones.

Before the skew beam is strengthened or replaced, temporary supports should be installed if necessary. When the skew beam is replaced, trigonal timber should be fabricated accurately and nailed firmly to the beam, while the nailing of step boards should be uniform. In addition, the upper and lower ends of the skew beams of stairs should be fixed firmly, and the regions close to the wall or on the ground should be treated to prevent corrosion.

When skew beams of hidden beam stairs are fabricated, the disposition of step slots in the skew beams should be accurate, steps should be connected to the beam firmly, the ends of the skew beams of stairs should be fixed firmly, and the connection of the skew beams should also be reliable.

When the ends of stairs are strengthened by splints, it is feasible to fabricate full-scale mould boards according to the dimensions of design and actual components, number the members individually, and manufacture strengthening components according to the mould boards.

5.5.3   Strengthening of Wooden Ceilings

Strengthening of wooden ceilings generally includes hanging ceiling on wooden purlins with wooden sag rods and using wooden grid to bear ceilings and installing insulating layer in cold areas.

Temporary support should be installed before wooden ceilings are strengthened.

The wooden sag rods splitting at ends should be replaced, while the number of wooden sag rods should be increased if there are not enough. The ends of each wooden sag rod should be pinned firmly by no less than two nails. When clamps are used in strengthening the damaged major wooden grid of ceiling (joist), it is feasible to fabricate full-scale mould boards according to the dimensions of design and actual components, and then manufacture strengthening components specifically for the mould boards. Nails (if any) should conform to relevant specifications.

5.6   Seismic Retrofitting of Wood Structure Buildings

5.6.1   Basic Principles of Seismic Strengthening

When wood structure buildings do not satisfy requirements after seismic assessment, seismic strengthening is required. Wood structures can better resist earthquakes. As long as the wooden components are free from decay, severe fracture, pull-out of tenons or inclination, and have connections to enclosure walls, damages of wood structure would be slight under earthquake action, even in high seismic intensity regions. Therefore, the key for seismic strengthening of wood structure buildings is their bearing systems, so the seismic capacities of wood frames should be enhanced. When conducting seismic strengthening, it is feasible to lighten roof weight, reinforce wood frames, strengthen and add braces, strengthen connections between wooden components or between enclosure walls and wooden components, install new seismic brick walls, and eliminate inadequate construction details.

When wood structure buildings are undergoing seismic strengthening, seismic checking calculations are not necessary. Section dimensions of the newly installed components for the seismic strengthening of wood structural buildings can be calculated from a static load. However, connections between the new and the original components need to be enhanced. Feasible methods and measures for seismic strengthening are adopted according to the actual situation.

Wood structure buildings, especially old wood frame buildings, which show decay, corrosion, erosion, deformation and cracks in separate positions such as beams, columns, roof trusses and purlins, should be strengthened.

5.6.2   Scopes and Methods of Strengthening

(1) Strengthening of wooden components such as beams, columns, roof trusses and purlins, etc.

a. To prevent horizontal displacement of roof skew beams or herringbone roof trusses in earthquake, it is feasible to strengthen them with steel tension rods (Fig. 5.19).

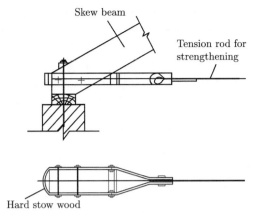

Fig. 5.19   Strengthening of skew beam with steel tension rods.

b. Wooden components such as roof trusses, beams, columns, which have cracks, decay, corrosion or erosion, can be strengthened according to non-seismic requirements.

c. When the ends of wooden beams have severe decay, the decayed parts can be cut off, and extended by channel steels as the new in-wall parts.

d. When the ends of roof trusses have severe decay, the decayed parts can be cut off and replaced by new components. If the decay cannot be eliminated, cut off the decayed region and replace the wooden joints with rolled steel components or reinforced concrete joints.

(2) Strengthening and adding supports or bracings

a. Vertical cross braces anchored by bolts can be added between trusses, especially at the ends of buildings. It will be beneficial to add stow woods at the intersections of cross braces, which will enhance the connection of cross braces, as shown in Fig. 5.20.

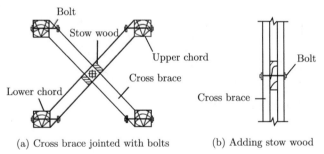

(a) Cross brace jointed with bolts   (b) Adding stow wood

Fig. 5.20   Conjunction of cross braces.

b. To improve the integral seismic capacity of roof trusses, transverse supports at the upper chord may be added, as shown in Fig. 5.21.

c. Braces should be added at the junctions of roof trusses and wooden columns. Joints of the braces are shown in Fig. 5.22. It will be beneficial to connect the braces with bolts, as shown in Fig. 5.23. Take wooden clamps as braces, and fix them with bolts; or take trigonal woods as stow woods, both of which function as braces, as shown in Fig. 5.24.

(3) Strengthening of the connection of wooden components

a. Install bolsters at the junctions of beams and columns, then anchor them with bolts to enhance the integrity. It is feasible to strengthen them according to non-seismic requirements, as shown in Fig. 5.6.

b. Utilize iron rod and bolt to connect roof truss and column as shown in Fig. 5.25.

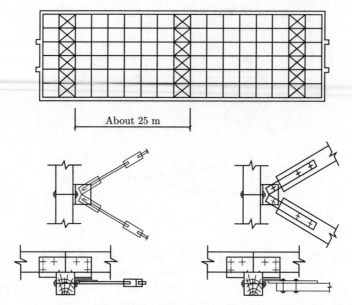

Fig. 5.21   Adding transverse bracing at upper chord of roof cushion plate.

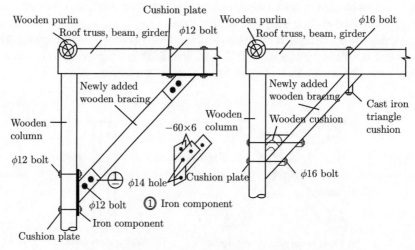

Fig. 5.22   Strengthening of wood frame with bracing (unit: mm).

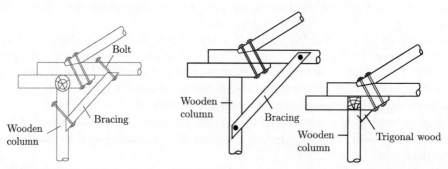

Fig. 5.23   Bracing joints with bolts.   Fig. 5.24   Wooden clamp or trigonal wood used as bracing.

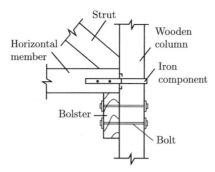

Fig. 5.25   Strengthening of joint of roof truss and column with bolts.

c. When wooden roof trusses and columns are connected by the tenon method, the trusses are likely to crack or even fracture due to weakened cross sections. It is feasible to use structural method shown in Fig. 5.26 to conduct strengthening.

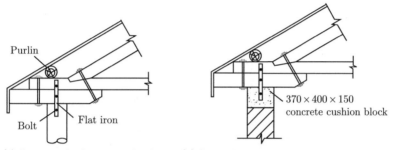

(a) Strengten by flat iron and bolt   (b) Strengthen by concrete cushion block and bolts

Fig. 5.26   Strengthening of column and overhang eaves (unit: mm).

d. Method of strengthening joints that connect the ends of wooden roof trusses and brick columns is shown in Fig. 5.27.

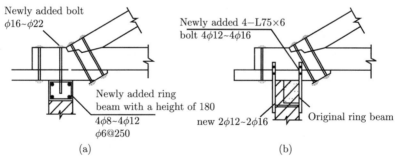

Fig. 5.27   Strengthening of supports of wooden roof trusses (unit: mm).

e. Fix the purlins firmly onto the roof trusses. It is feasible to connect the purlins to the roof truss with nails by making dovetail grooves onto the purlins, or with short battens.

(4) Supporting length of wooden roof trusses or wooden beams

When the supporting length is less than 250 mm and there are no anchoring measures, it is feasible to adopt the following methods to strengthen the supports:

a. Adhere wooden columns or build masonry columns at top.

b. Install new bolsters along the inside surfaces of brick walls and wooden clamps to extend the supports, as shown in Fig. 5.28 and Fig. 5.29.

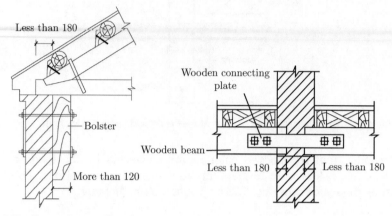

Fig. 5.28   Strengthening of roof
truss with bolster (unit: mm).

Fig. 5.29   Extension of support with
wooden clamp (unit: mm).

c. Take new anchoring measures at roof supports, as shown in Fig. 5.30.

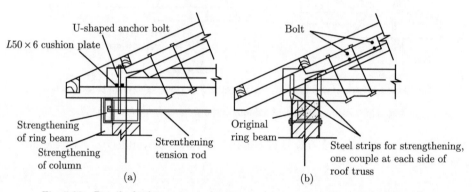

Fig. 5.30   Detail of taking new anchoring measurements at support (unit: mm).

(5) Connections of wood frames and walls

a. For strengthening of walls and wooden beams or wooden joists, connection between walls and wooden beams is shown in Fig. 5.31(a), and connection between walls and wooden joists by wall tenon is shown in Fig. 5.31(b).

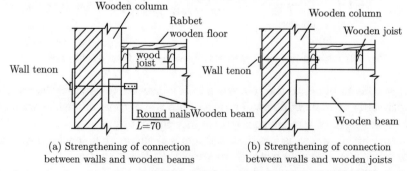

(a) Strengthening of connection
between walls and wooden beams

(b) Strengthening of connection
between walls and wooden joists

Fig. 5.31   Strengthening of connection between walls and wooden beams or wooden joists (unit: mm).

b. Strengthening of connection between post-masonry partition walls and wooden columns, girders or beams.

Post-masonry partition walls of 120 mm in thickness and over 2.5 m in height, or of 240 mm in thickness and over 3 m in height, should be connected to wood frames with a course of rebar ($2\phi6$, 700 mm long) every 1 m along the wall.

c. Strengthening of connection between walls and corner columns is shown in Fig. 5.32.

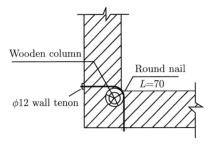

Fig. 5.32   Strengthening of connection between walls and corner columns (unit: mm).

Index

For Product Safety Concerns and Information please contact our
EU representative GPSR@taylorandfrancis.com Taylor & Francis
Verlag GmbH, Kaufingerstraße 24, 80331 München, Germany